Apple Pro Training Series

Final Cut Server 1.5

Drew Tucker

Apple Pro Training Series: Final Cut Server 1.5
Drew Tucker

Published by Peachpit Press. For information on Peachpit Press books, contact:
Peachpit Press
1249 Eighth Street
Berkeley, CA 94710
(510) 524-2178
www.peachpit.com
To report errors, please send a note to errata@peachpit.com. Peachpit Press is a division of Pearson Education.

Project Editor: Nancy Peterson
Development Editor: Stephen Nathans-Kelly
Production Coordinator: Happenstance Type-O-Rama
Contributing Writers: Brendan Boykin and Matt McManus
Technical Editors: Matthew Geller
Copy Editors: Darren Meiss, Elissa Rabellino, and Karen Seriguchi
Media Reviewer: Jay Payne
Media Producer: Eric Geoffroy
Compositor: Chris Gillespie, Happenstance Type-O-Rama
Indexer: Joy Dean Lee
Cover Illustrator: Kent Oberheu
Cover Producer: Happenstance Type-O-Rama

ISBN 13: 978-0-321-64765-8 ISBN 10: 0-321-64765-3
9 8 7 6 5 4 3 2 1 Printed and bound in the United States of America

Contents at a Glance

Table of Contents

USER ▶ Signifies a section primarily for users.
ADMIN ▶ Signifies a section primarily for administrators.

Getting Started

Welcome to the official Apple Pro Training course for Final Cut Server, the powerful media asset management and workflow collaboration software from Apple. This book includes exercises and media files to help you learn the various ways Final Cut Server can help your workflow.

The Methodology

This book presents various exercises for both administrators and users to experience the features of Final Cut Server. To get the most from this book, everyone should do all of the exercises, as the administrator parts are required to configure the system for the user parts.

Course Structure

This book's 11 lessons take you through the lifecycle of an asset within Final Cut Server. Within each lesson, the exercises designated "Admin" or "User" signify who would be performing this task on an active Final Cut Server system outside the classroom: an administrator or a user. (Exercises that have neither designation are equally relevant to each group.)

Overview and Installation

Lesson 1 introduces you to Final Cut Server: the terms and the high-level workflow. Then it walks you through installing Final Cut Server and the client Java application.

Ingesting Assets

Lessons 2 through 4 cover the various methods of bringing files into Final Cut Server. These lessons lead you through both the manual and automated methods of making files into Final Cut Server assets.

Working with Assets

Lessons 5 through 8 use your ingested assets to explore how Final Cut Server can maximize your workflow, even when dealing with thousand of assets. Beyond just searching and organizing, these lessons show you the Final Cut Server features for collaborating with other users and other applications.

Publishing and Archiving

Lessons 9 and 10 guide you through delivering your assets via manual and automated processes. Once the assets are delivered, this section covers how to archive assets to create room for new assets while continuing to track, preview, and restore older assets.

Advanced Administration

Lesson 11 is, in essence, an administrator's showcase. Here Final Cut Server administrators will find additional features that will improve the workflow of users and help keep the database healthy.

System Requirements

For this curriculum, you will be running the Final Cut Server software and client application on the same workstation. As such, the minimum hardware and software required is that of the server minimum requirements:

- An Intel-based Mac computer
- 2 GB of RAM
- ATI or NVIDIA graphics processor (integrated Intel graphics processors not supported)
- Mac OS X 10.5.6 or later
- QuickTime 7.6 or later
- DVD drive
- 1.5 GB of disk space for the applications
- 4.2 GB of disk space for the curriculum media
- Final Cut Studio (2009) applications with the latest software updates applied

Recommended system requirements are constantly evolving as new configurations of hardware and software become available. To check for the latest requirements, go to www.apple.com/finalcutserver/specs.

Before beginning to use *Apple Pro Training Series: Final Cut Server 1.5*, you should have a working knowledge of your Mac and its operating system. Make sure that you know how to use the mouse and standard menus and commands, and also how to open, save, and close files. If you need to review these techniques, see documentation included with your Mac.

Preparing the Workstation

To follow along with all the lessons, you'll need to configure your workstation and copy the lesson files onto your hard drive. The lessons are located on the DVD accompanying this book. You'll need approximately 4 GB of free disk space on your hard drive.

The exercises in this book require administrator privileges on your workstation. This curriculum was written for the classroom environment; please exercise caution if you're using these exercises in an active post-production environment.

1 From the Apple menu, open System Preferences and choose the Accounts pane.

2 If you've not done this already, create an Administrator account with a Full Name of Administrator. For Account name, use *admin* with a password of *admin*.

NOTE ▶ This account is for training purposes only and should never be used in a real-world environment.

3 Quit System Preferences.

4 Insert the APTS_FinalCutServer DVD into your computer's DVD drive.

   The Finder window opens, displaying the contents of the DVD. If the window doesn't appear, double-click the APTS_FinalCutServer DVD icon to open it.

5 Drag the FCS_Book_Files folder from the DVD onto the administrator's desktop.

   The files are copied onto your hard drive.

NOTE ▶ The location of this folder is important. Make sure this folder resides on the desktop so that the book files are easy to find via the file paths provided as you work through the exercises in the book.

6 Once the FCS_Book_Files folder is copied to your hard drive, you may eject the DVD.

About the Apple Pro Training Series

Apple Pro Training Series: Final Cut Server 1.5 is both a self-paced learning tool and the official curriculum of the Apple Pro Training and Certification Program.

Developed by experts in the field and certified by Apple, the series is used by Apple Authorized Training Centers worldwide and offers complete training in all Apple Pro products. The lessons are designed to let you learn at your own pace. Each lesson concludes with review questions and answers summarizing what you've learned, which can be used to help you prepare for the Apple Pro Certification Exam.

For a complete list of Apple Pro Training Series books, see the ad at the back of this book, or visit www.peachpit.com/apts.

Apple Pro Certification Program

The Apple Pro Training and Certification Programs are designed to keep you at the forefront of Apple digital media technology while giving you a competitive edge in today's ever-changing job market. Whether you're an editor, graphic designer, sound designer, special effects artist, or teacher, these training tools are meant to help you expand your skills.

Upon completing the course material in this book, you can become an Apple Certified Pro by taking the certification exam online or at an Apple Authorized Training Center. Certification is offered in Final Cut Pro, Final Cut Server, Color, Compressor, Motion, Soundtrack Pro, DVD Studio Pro, and Logic Pro. Certification as an Apple Pro gives you official recognition of your knowledge of Apple professional applications while allowing you to market yourself to employers and clients as a skilled, pro-level user of Apple products.

For those who prefer to learn in an instructor-led setting, Apple offers training courses at Apple Authorized Training Centers worldwide. These courses, which use the Apple Pro Training Series books as their curriculum, are taught by Apple Certified Trainers and balance concepts and lectures with hands-on labs and exercises. Apple Authorized Training Centers have been carefully selected and have met the highest standards in all areas, including facilities, instructors, course delivery, and infrastructure. The goal of the program is to offer Apple customers, from beginners to the most seasoned professionals, the highest-quality training experience.

For more information, please see the ad at the back of this book, or to find an Authorized Training Center near you, go to training.apple.com.

Companion Webpage

As Final Cut Server 1.5 is updated, Peachpit may choose to update lessons or post additional exercises as necessary on this book's companion webpage. Please check www.peachpit.com/apts.fcs1.5 for revised lessons or additional information.

Resources

Apple Pro Training Series: Final Cut Server 1.5 is not intended as a comprehensive reference manual, nor does it replace the documentation that comes with the application. For comprehensive information about program features, refer to these resources:

- The *Final Cut Server Setup* and *Administrator Guides*: Available at http://documentation.apple.com, these guides contain a complete description of all features.
- Apple website: www.apple.com.

1

Goals

Understand Final Cut Server terms

Install Final Cut Server

Install the client application

Tour the client application interface

Lesson **1**

Overview and Installation

With Final Cut Server, Apple provides an essential and long-sought-after solution for the collaborative requirements of digital creative work. Final Cut Server features a powerful database and an easy-to-use interface that allow it to meet three important needs:

- It manages assets by interacting with both creative application software—most notably Final Cut Pro—and an organization's networked storage systems.
- For users inside or outside an organization, it acts as a conduit to collaborative steps such as reviewing shots, annotating sequences, and checking files in and out.
- It automates workflow as it generates proxy (low-resolution) versions of clips for quick access; copies and transcodes assets from one storage system to another; scans for new and changed media; and looks for changes in asset metadata, triggering email notifications and a host of other automated actions.

The Nature of the Software: Server, Client, and Device

Final Cut Server gets its name from its nature: It's *server* software that runs on a Mac somewhere in your organization. However, the main way that you interact with the software is through a *client* application on any other computer, inside or outside the organization. This client application was developed in the Java programming language and can run on a Mac or PC that is connected to your network.

You'll use the client application to work with all of the files within Final Cut Server, including placing new files inside it and transferring files from it into the applications you use often, such as Final Cut Pro.

Final Cut Server connects to any number of *devices* containing media and other files that are important to your workflow. For Final Cut Server to work properly, any place where your organization stores files should be defined as a device. Here are some device examples:

- An Xsan volume
- A network-accessible file server
- Hard drives that are directly connected to Final Cut Server

The files on these devices are accessible through the network that connects your computer to Final Cut Server. The speed of that connection, whether it is contained within your organization or handled remotely by a virtual private network (VPN), will determine how fast you can access these files.

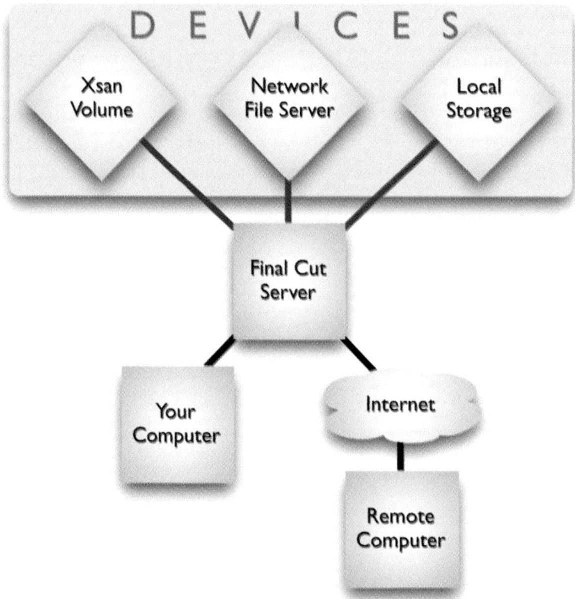

If your computer and Final Cut Server are Xsan clients accessing the same Xsan volume, Final Cut Server will simply provide pointers to the files on that volume, enabling you to access the files instantaneously. Even in this case, Final Cut Server can also act as a gateway to other devices in your organization.

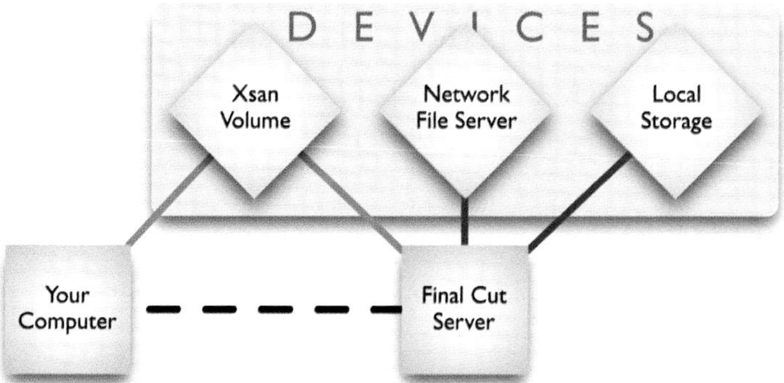

Naming Conventions

To avoid confusion when referring to the various manifestations of Final Cut Server, the following terms will be used from now on:

- *Final Cut Server,* used alone, refers to the Mac that is running installed and configured Final Cut Server software.

  It may also refer to the interaction between the Final Cut Server and the client application in general.
- *Final Cut Server software* refers to the software that gets installed and configured on the Mac that will become the Final Cut Server.
- *Final Cut Server client application,* or just *client application,* refers to the Java client software that users download from the Final Cut Server to install and run on any computer that wishes to communicate with the Final Cut Server.

The Building Blocks: Assets, Proxies, Metadata, Productions, and Jobs

Final Cut Server contains a central catalog of *assets*, which are references to actual media files located on a device. Each asset contains pointers to the original high-resolution version of the file, which we'll call its *primary representation.*

When an asset is added to the Final Cut Server catalog, it makes a *proxy* of the primary representation—a much smaller, lower-quality (and therefore easier-to-transmit) version of the primary representation that is usually stored on the hard disk(s) of Final Cut Server. Users that connect to Final Cut Server using low-bandwidth connections will view and use proxy files in order to save transmission time and hard disk space.

Each asset contains a *thumbnail* image of the primary representation for quick identification. The asset also contains a rich assortment of *metadata* (descriptive data about data), which helps you in searching for the asset, using it in your organization's workflow, and categorizing it for special tasks such as archiving or deletion. Some metadata is generated automatically when the asset is created, but a lot of metadata is entered by you and your colleagues in order to document the asset's attributes and usage.

Assets can be pooled together in a group called a *production.* A production is like a virtual folder inside Final Cut Server: a group of assets being used for the same purpose. Final Cut Server can assign metadata to productions automatically, and users can add metadata to them. Productions have their own catalog within Final Cut Server that can be searched and categorized.

Final Cut Server also tracks *jobs*—the actions that you, your colleagues, and Final Cut Server itself perform, such as adding assets to a catalog or copying primary representations from one device to another. All jobs are logged and can be searched by users. Jobs have certain metadata attached to them, including error messages that can be looked up in case a job fails.

Collaboration: Users and Groups

Obviously, Final Cut Server will be accessed by *users,* people inside and perhaps outside the organization. In order to gain access to the catalogs within Final Cut Server, you'll log in to it with a user name and a password. This provides security for assets and productions that have sensitive materials. The Final Cut Server administrator gathers certain users together into *groups.* These groups will have certain access and functionality privileges within Final Cut Server, so that everyone has access to the appropriate assets and functions.

ADMIN ▼

Installing Final Cut Server

The Final Cut Server software is very easy to install. This exercise steps you through a basic installation of the Final Cut Server software for use in a video post-production facility.

NOTE ▶ This scenario is designed for classroom use, so the server software and client application will be accessed on the same machine. The devices used for assets and proxies will also be living on the same machine. Please refer to the *Final Cut Server Setup* and *Administrator Guides* for more detail regarding installation beyond the classroom.

1 Insert your Final Cut Server Install DVD into your computer or double-click the Final Cut Server Install.dmg identified by your instructor.

2 Double-click the Install Final Cut Server icon. Then, if a security message appears, click Continue.

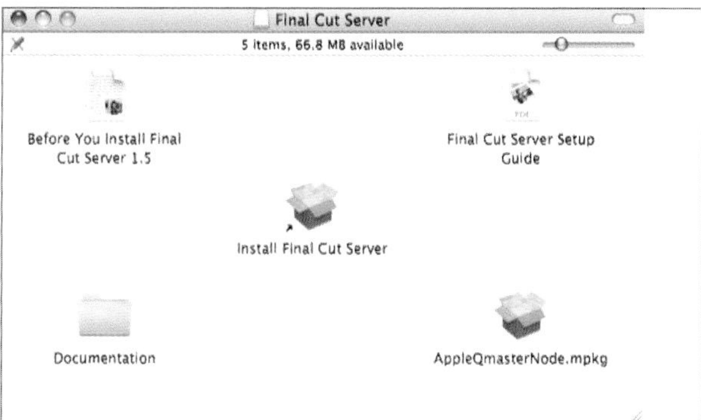

3 In the "Welcome to the Final Cut Server Installer" pane, click Continue.

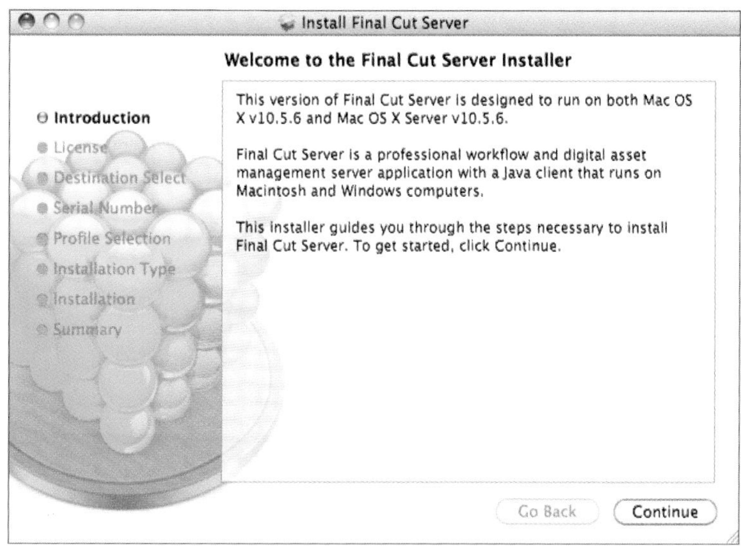

4 Click Continue to the Software License Agreement, and then click Agree.

5 Enter your first and last name, organization (if desired), followed by the Final Cut Server software's serial number provided by your instructor. Click Continue.

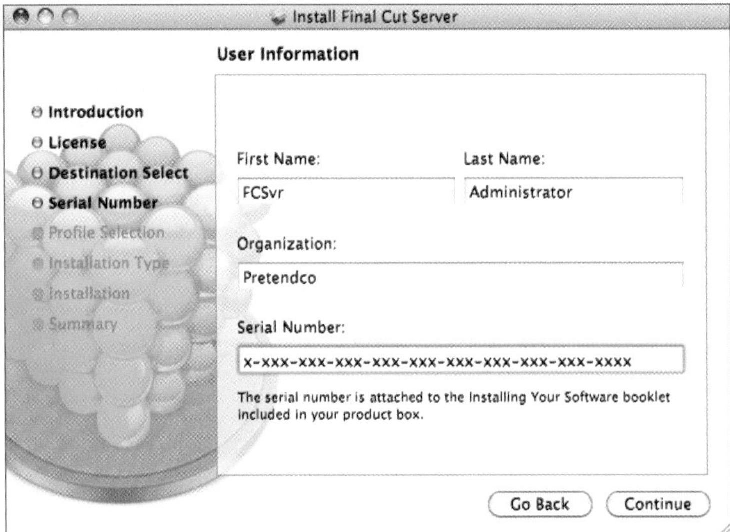

The Customer Profile Selection pane creates the Final Cut Server database using preset metadata structures. These presets give you a great launching point for how to organize your assets through metadata sets and workflows specific to each customer profile.

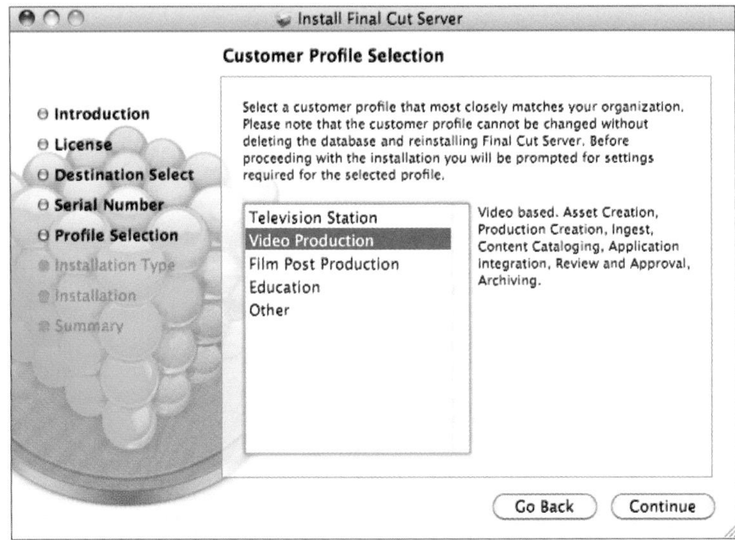

NOTE ► Although each profile is customizable once the database is created, the profile itself cannot be switched.

6 Select the Video Production profile and click Continue.

In the "Settings for Profile" pane, you configure initial device locations for assets and their various *proxies*—in other words, where your representations will live. This pane also has settings that begin configuring the workflow automation features of Final Cut Server.

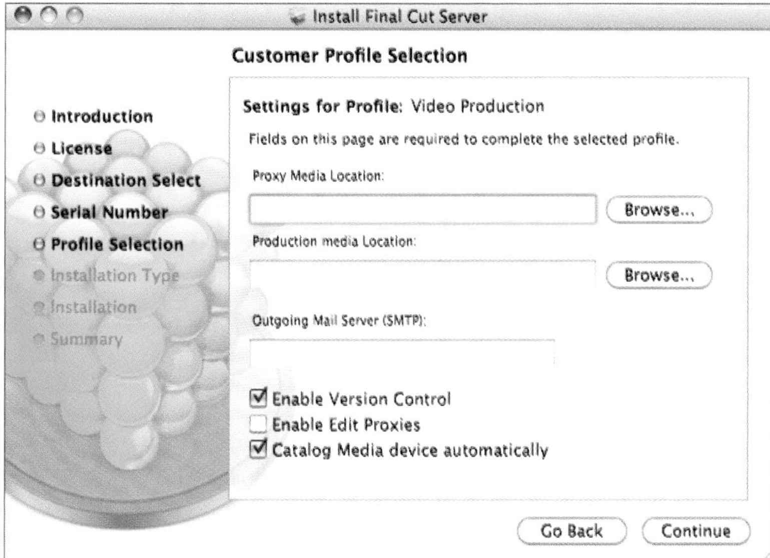

For this part of the exercise, you'll create a folder called *FCSvr* at the root level of the system hard drive for use of all representations beginning with the proxies. You will then assign that folder for use within this pane.

7 To the right of the Proxy Media Location field, click Browse. In the Finder dialog that appears, navigate to your system drive (Macintosh HD) in the left sidebar. Click New Folder and rename the folder *FCSvr*. With the new folder FCSvr selected, click Open.

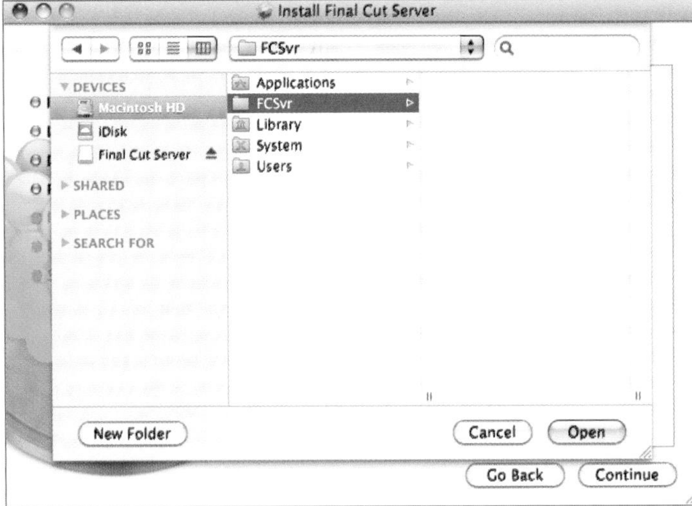

The Proxy Media Location field will not be populated as shown in the following figure. A variety of proxies created for new assets will reside here.

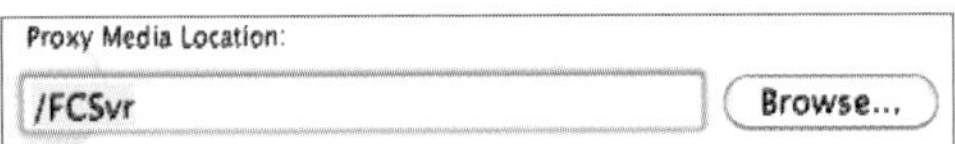

NOTE ▶ Refer to the *Final Cut Server Setup* and *Administrator Guides* for recommendations on placement of the proxy files.

Now you will select the new FCSvr folder as the destination for primary representations, edit proxies, and versions.

8 To the right of the Production Media Location field, click Browse. In the Finder dialog, ensure that the FCSvr folder is selected, and then click Open.

NOTE ▶ In a typical installation, these two devices will be set to different volumes. The Production Media Location should be large, expandable, fast, and redundant. Although the standard proxies (not including edit proxies) are relatively small compared to primary representations, the proxies device will get a workout. All users will be constantly accessing this device. Fast and redundant with room to grow is desired here as well.

9 If provided by your instructor, enter the non-authenticating SMTP server for the network into the Outgoing Mail Server field; otherwise, leave it blank.

The next three selections will vary from facility to facility outside the classroom. With Enable Version Control selected, Final Cut Server will automatically create and track changes to assets that are checked in or uploaded. In essence, version control creates a "history" of an asset.

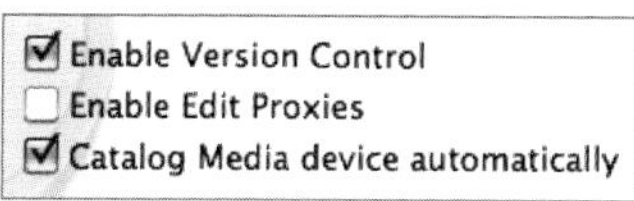

10 If necessary, select Enable Version Control, which is selected by default.

The Enable Edit Proxies option creates "offline quality" media files of an uploaded Final Cut Pro project. Typically, this function is used to create smaller, editable media files for an offline/online edit workflow. The quality (and size) of the resulting proxies can be controlled within the client application's Administration window.

11 Select the Enable Edit Proxies option.

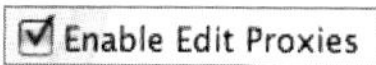

The "Catalog Media device automatically" option creates one of the many automated workflow features of Final Cut Server. If selected, it creates a scheduled automation that scans the production media device for changes and adjusts the Final Cut Server database accordingly. You may choose not to enable this feature during installation but retain the option to configure later.

12 Deselect "Catalog Media device automatically." You will create this manually in a later lesson.

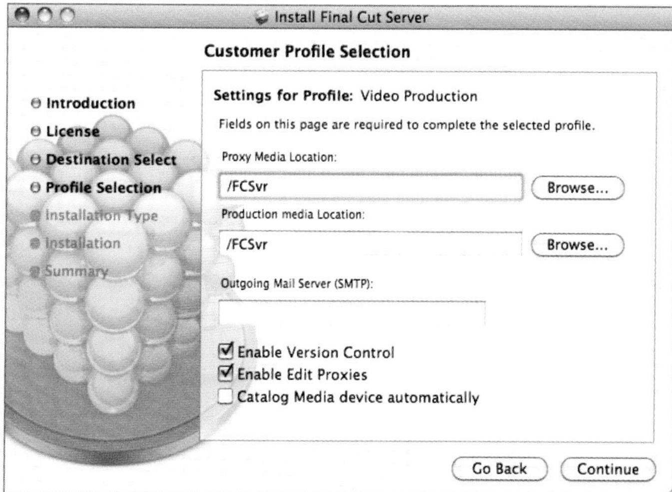

13 Click Continue. When a new pane opens, click Install.

14 Authenticate the installer with the username and password provided by your instructor.

15 When installation has successfully completed, click Close.

16 Eject the install DVD or Install.dmg file.

USER ▼

Downloading and Installing the Client Application

The requirements for the Final Cut Server client application are simple: It must run on a Mac OS X, Windows XP SP2, or Windows Vista system on which a runtime version of Java and QuickTime are installed. Java and QuickTime come preinstalled on Mac OS X. Some PCs may require the installation of Java and QuickTime before you can install the Final Cut Server client application.

To install the client application, you download the application from the Final Cut Server. The application is available for download via a web browser at the network address, or URL, of the Final Cut Server.

The download URL can be either a specific domain name or an IP address, followed by */finalcutserver*, such as the following:

http://pretendco.com/finalcutserver

http://10.0.0.2/finalcutserver

NOTE ▶ This webpage is only for downloading the Final Cut Server client application. Once you have downloaded and installed the client application, you will start it by opening an application on your own computer.

1 Open Safari and enter the URL for Final Cut Server: *http://localhost/finalcutserver*.

When you go to this URL, you will first see a page that prompts you to download the Final Cut Server client application.

2 Click Download. The client application install begins. Don't click anything; the installer will proceed automatically.

NOTE ▶ The application downloads slightly differently on a Mac versus a PC. Refer to the *Final Cut Server Setup* and *Administrator Guides* for information on installing the client application on a Windows system.

After you click Download, you'll see a progress indicator window showing that the software is downloading.

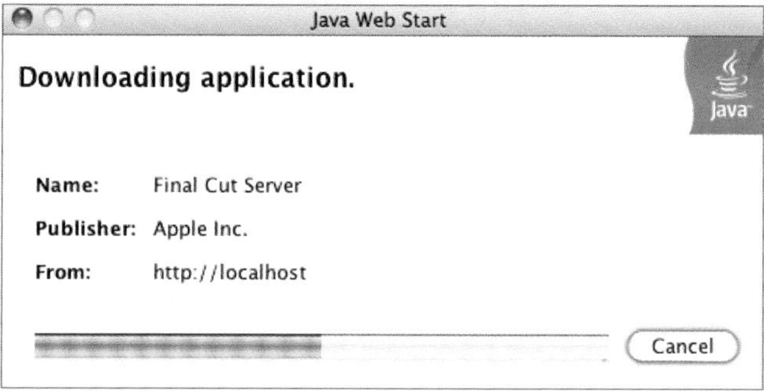

When the download is complete, continue with the following steps.

3 In the window that appears, select "Allow all applications from 'localhost' with this signature," and then click Allow.

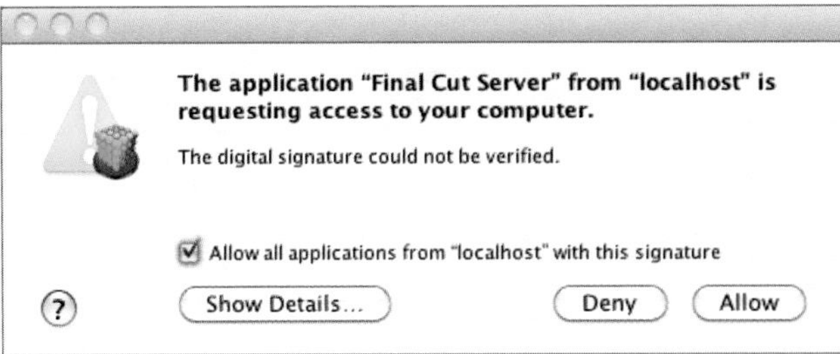

By clicking Allow, you're identifying the Final Cut Server as a trusted source. Every time you launch the client application, a software update is performed against the Final Cut Server. If a newer version of the client application is available, it will be downloaded.

In the next window, you'll choose where to save the client application. The default location is the desktop. Once Final Cut Server is installed, you'll no longer need to use the Finder to find your media files. That's the whole point of Final Cut Server: an easier way to find your media assets.

4 Leave the default name in the Save As field, and leave the desktop as the default location for the client application. Click Save.

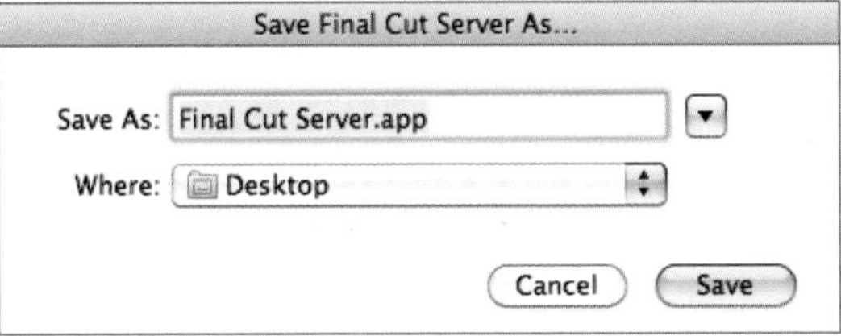

NOTE ► From now on, you can simply double-click the Final Cut Server application shortcut on your desktop to access the application.

The installation continues automatically; it may take some time. When the installation is complete, the client application will open and display the software license agreement.

5 Click Agree to accept the terms and conditions for the client application.

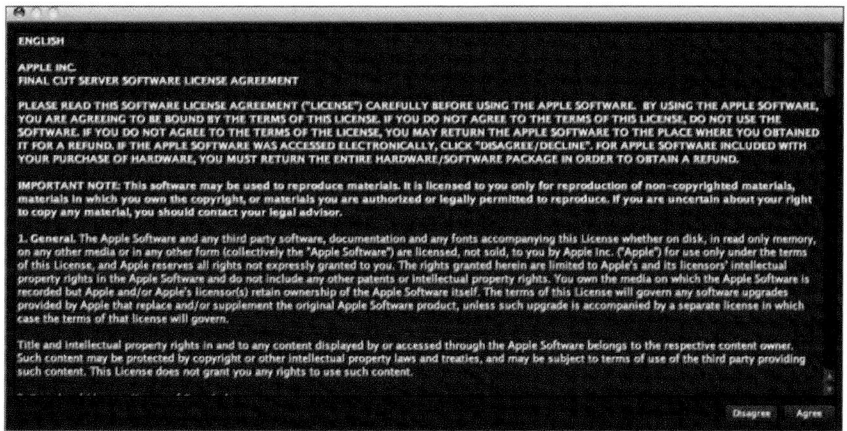

When the installation process is complete, you'll see the Login window. When you open Final Cut Server in the future, the Login window will appear, as shown in the next exercise.

USER ▼

Logging In to Final Cut Server

The credentials used to log in to Final Cut Server are determined by your network's setup. These might be unique, or they may be the same ones you use for other services at your organization. Your username is grouped with other similar users to determine your permissions within Final Cut Server. Assigning groups and permission sets is covered later in this book.

> **NOTE ►** Final Cut Server may be configured to use Local User group mapping, Open Directory, or Active Directory. This book uses locally configured users and groups. An administrator user assigned to the default admin permission set will be used initially.

1 In the Final Cut Server Login window, enter the username and password provided by your instructor.

Logging in brings you to the main interface. Currently, your Assets and Productions panes will be empty because you just installed Final Cut Server. In Lesson 2 you'll catalog assets using the Upload process.

Window pop-up menu

Productions pane

Server pop-up menu

Assets pane

Toolbar

Search field

Smart Searches

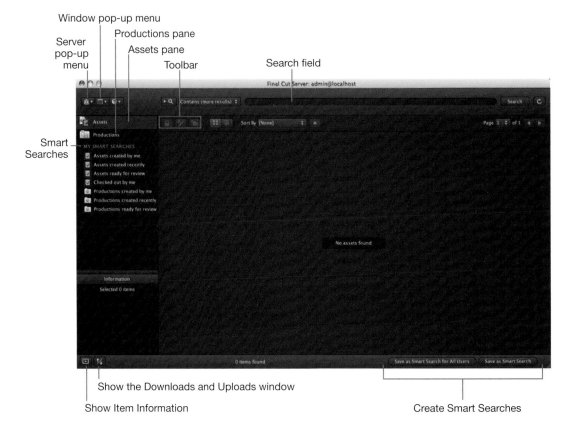

Show the Downloads and Uploads window

Show Item Information

Create Smart Searches

The key elements of the main Final Cut client application give you access to many features:

- Three pop-up menus (the Server, Window, and Help menus) provide extra functions, such as access to preferences and Final Cut Server documentation.
- Clicking the Assets button will show you the assets you have, while clicking Productions will show you the assets you have grouped into virtual folders.
- The Toolbar contains buttons for frequently used functions such as locking assets.
- The search field is where you can enter keywords or more advanced metadata to search for assets or productions.
- Clicking the Show Item Information and Downloads and Uploads buttons reveals additional metadata and job status information.
- The Smart Searches list shows favorite searches for assets and productions.

In the next lessons, you will learn how to work with all the features of the Final Cut Server client application.

Lesson Review

1. What is the difference between a primary representation and a proxy?
2. What are the two main panes within the Final Cut Server client application?
3. What is metadata?
4. Where can you access the Final Cut Server Preferences window within the client?

Answers

1. The primary representation is the original file from which the asset was made. Proxies are smaller, lower-resolution files derived from the primary representation of a video asset. They're more easily transmitted, especially over a low-bandwidth connection.
2. The Assets and Productions panes, which can be accessed from the main client application window in the upper-left corner.
3. *Metadata* means data about data. It is used to describe, find, and repurpose content. The Final Cut Server client application has powerful search fields where you can search for assets and productions using metadata information.
4. Preferences can be found in the Server pop-up menu in the upper-left corner of the main client window.

2

Goals

Discover the structure of metadata

Upload assets into Final Cut Server

Lesson **2**

Creating Assets via Upload

USER ▶ This lesson is primarily for the user, although admins will also find the information useful.

The entire creative process hinges upon you and others in your organization creating media files. Whether those are video files from a camera, still images from a DSLR, music files from Logic Pro, or even PDFs of a script, these media files are useless if you and your colleagues can't find them. The way to ensure that everyone in your organization can find these files is to create them as assets within Final Cut Server with appropriate metadata. The most basic method of creating these assets is uploading.

As you learned in Lesson 1, when you upload files to Final Cut Server, a series of representations is created for each asset. These representations enable any user to preview an asset. More importantly, these representations are stored so that anyone with permission can access them. Final Cut Server becomes a gateway to your files so that your colleagues won't call you on your day off as they rummage through your system looking for the file they need.

This lesson covers the upload method for creating assets. Subsequent lessons cover other methods of creating assets both from automation and from Final Cut Pro projects.

Introducing the Structure of Metadata

A key thing to remember when uploading media is metadata entry. Metadata is essential to finding assets within Final Cut Server. The more metadata you have (and the more accurate that metadata is), the easier it will be for you and your colleagues to find desired assets when searching through the thousands and thousands of assets inside Final Cut Server. To organize that search, metadata is ordered into *fields*, *groups*, and *sets*.

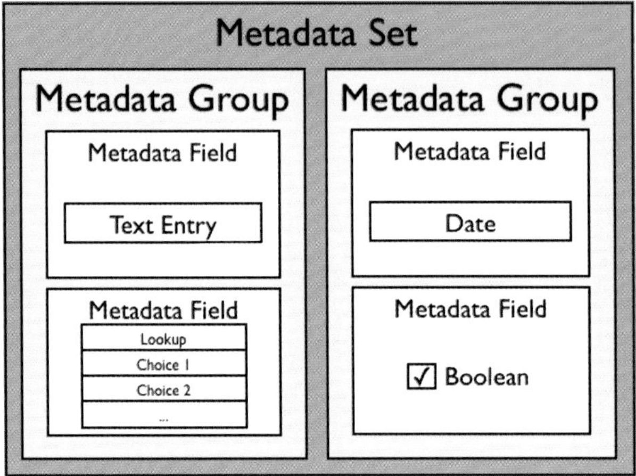

Basic metadata starts as a field. One metadata field that is always required is the filename. Final Cut Server populates the filename metadata field with the existing filename. It will also help you fill in the blanks with any metadata that it recognizes from a file. However, there are usually plenty of other blanks for you to fill in, such as *lookups*. Lookups are preset pop-ups from which you choose the appropriate option. The benefit of using lookups is not only their ease of use, but also that they help those of us who don't use a spell-checker.

The number of potential metadata fields can seem overwhelming, so metadata fields are gathered into logical families called metadata groups. Depending on the needs of your organization, you may use the default groups or choose among many customized groups.

Lastly, the groups are arranged into metadata sets. Some sets contain the same groups; some sets are completely different. Again, this is to simplify and specify the number and types of metadata fields for an asset.

NOTE ▸ Once a metadata set is applied to an asset, the chosen set cannot be changed. The metadata entered into the fields can be altered, but not the choice of which metadata fields/groups are used.

Manually Adding Assets

Uploading is a manual process. Whether you are uploading one asset at a time or a batch of files at once, you must initiate the process. The upload method is used for three reasons:

- To copy newly created files from your computer to a device that is connected to Final Cut Server or to reference newly created files that are already on a device
- To create assets for the files, which will automatically generate other representations, including clip proxies for video files
- To have an opportunity to customize the metadata for the assets that will be created for the files

You can use two methods for uploading files:

- Choose Upload File from the Server pop-up menu.
- Drag and drop.

NOTE ▶ Because the client application communicates with Final Cut Server over a network connection, you can add assets to your organization's devices no matter where you are. Be aware, however, that the speed of your connection to Final Cut Server will determine the speed of the upload of the files.

Using the Upload Files Command

Using the Upload Files command is the best method of uploading a file if your file is not handily available for drag-and-drop upload. This method will perform the upload similar to using an Import dialog within Final Cut Studio.

1 From the Server pop-up menu in the upper left of the client application window, choose Upload File.

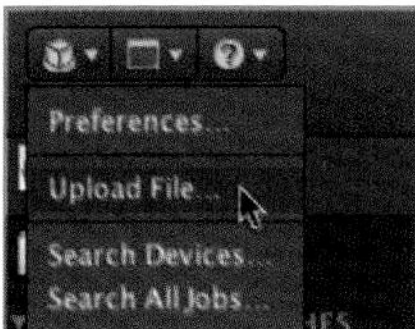

An Upload dialog appears, prompting you to choose a file, a series of files, or a folder.

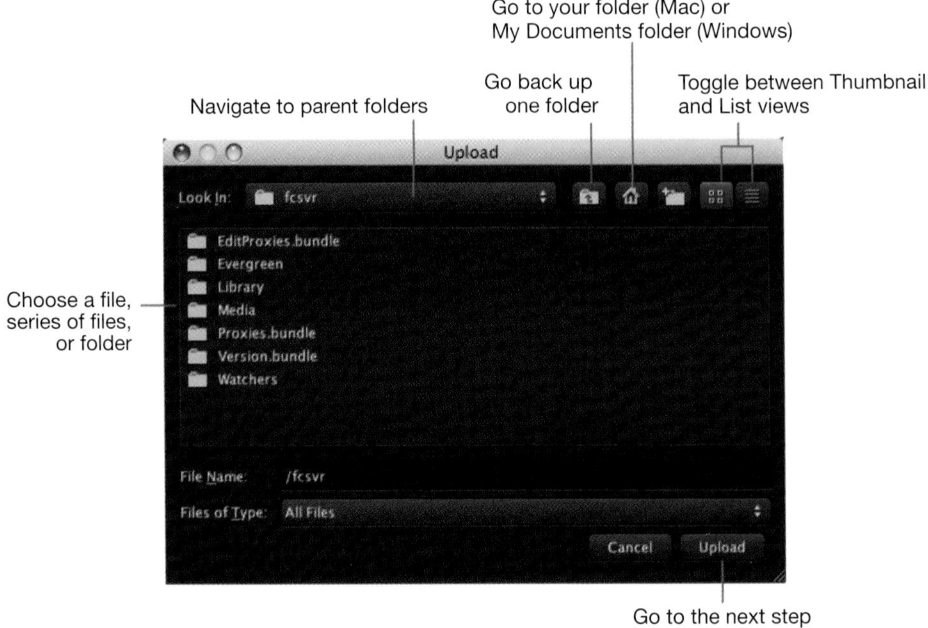

2 Navigate to your desktop by double-clicking the Desktop folder.

3 Double-click the FCS_Book_Files folder and then the For Upload folder. Select the **SR1020 015 Glacier Island-v** file, and then click Upload.

The Upload window appears, prompting you for several important pieces of information:

- The name of the copied file, usually kept as is. (This field appears only when you are uploading a single file.)
- The device, and possibly a path on that device, where you want the file to be placed so that Final Cut Server can manage it.
- Whether the asset will be associated with a *production*—a virtual folder for organizing your assets.

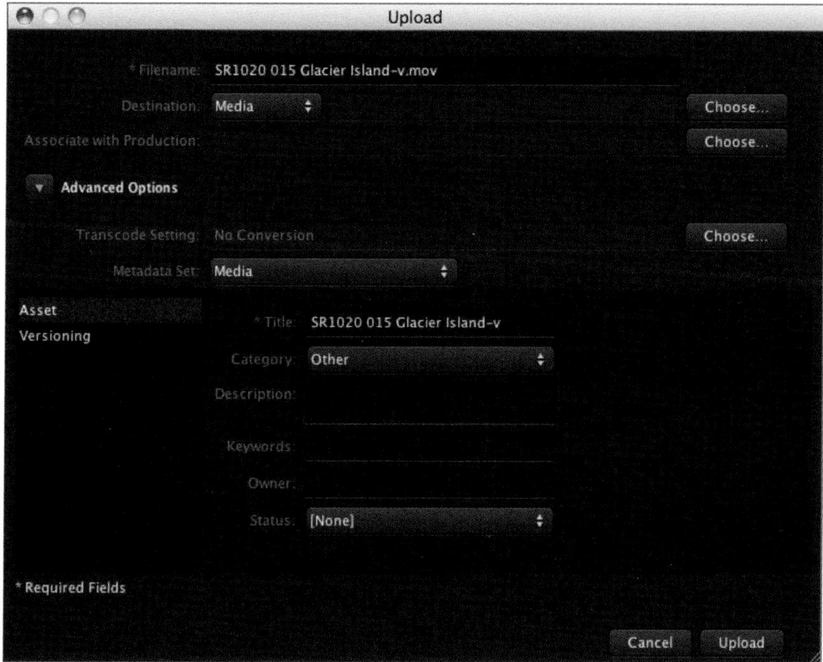

4 Leave the filename as is.

5 From the Destination pop-up menu, choose the Media device.

NOTE ▶ It's important to know beforehand which device (and possibly a subfolder) should receive your uploaded files, since every device varies in speed and capacity. There are also specialized devices, such as Library and Media, that might be appropriate. Also note that the device to be used may change from time to time or project to project. All users should check with the Final Cut Server administrator about the best device to store your uploaded files.

TIP If necessary, once your device is selected, click Choose to specify the subfolder path for the file.

You will not associate this asset with a production. You'll learn more about productions in Lesson 5.

6 In the Upload window, click the Advanced Options disclosure triangle, if necessary, to reveal additional options.

These advanced options are where you enter additional metadata about the new asset. Those metadata fields marked with an asterisk next to their names are required. An administrator may set additional fields as required before starting an upload. Take just a few moments to accurately complete the metadata fields so that when you go searching for an asset later, you can quickly find it.

Begin assigning metadata by choosing the appropriate metadata set. In this exercise, you're uploading a video file, so you will choose Media. Other preset options include Audio, Document, Graphic, and Project.

7 For Metadata Set, choose Media.

The default Media metadata set includes two metadata groups—Asset and Versioning—when uploading. The Asset group is where you will enter additional metadata for this asset.

8 With the Asset metadata group selected, leave the Category as Other.

Category is an example of a lookup. An administrator may customize this list to fit your workflow.

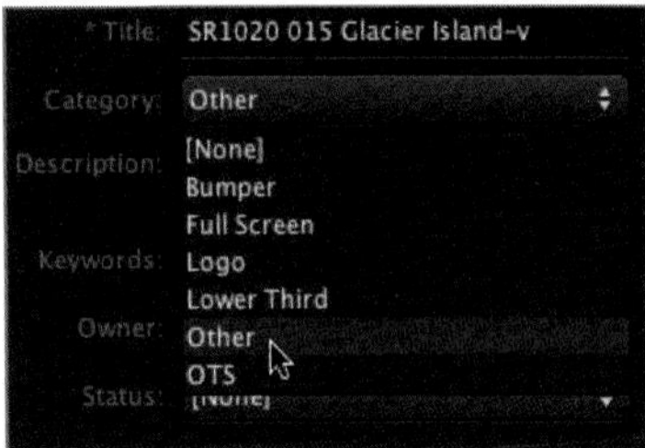

9 For Description, type *Flyover of Glacier Island*. Press Control-Tab to advance to the next field.

10 For Keywords, type *Aerial Sea Trees Glaciers Snow*. Press Control-Tab to advance to the next field.

11 Enter your name as the Owner.

12 From the Status lookup, choose New.

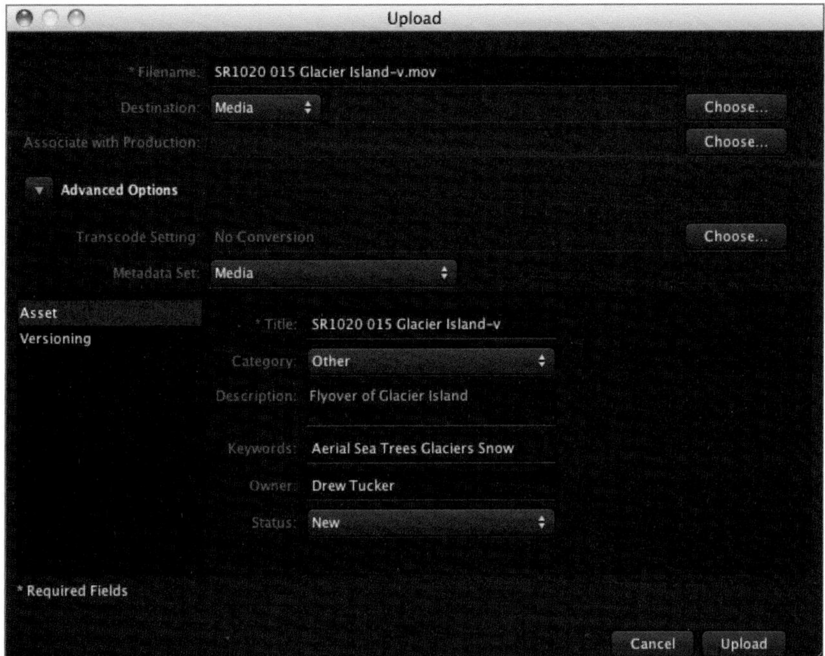

13 Finally, click Upload to start the process of uploading the file to Final Cut Server. You can view the progress of the upload by clicking the Downloads and Uploads button in the lower left of the client application window.

As you become comfortable with the upload process, you can simply watch the Jobs in Progress notification on the left side of the interface. This notification will stay visible during the upload as the primary and proxy representations are being created and stored within the designated devices.

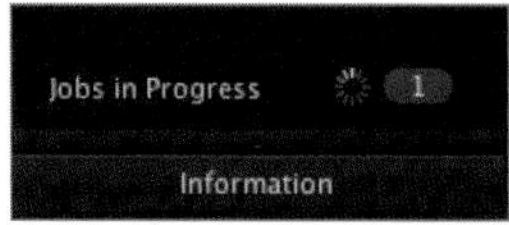

NOTE ▶ When the upload is complete, you will notice that the middle of the Assets pane still says "No assets found." Final Cut Server does not push new asset notifications to the client application. You would find the constant refreshing annoying as other users are modifying assets. To manually refresh the catalog display of assets, click the Refresh or Search buttons at the top right of the interface.

Now your Assets pane updates to show the newly uploaded asset.

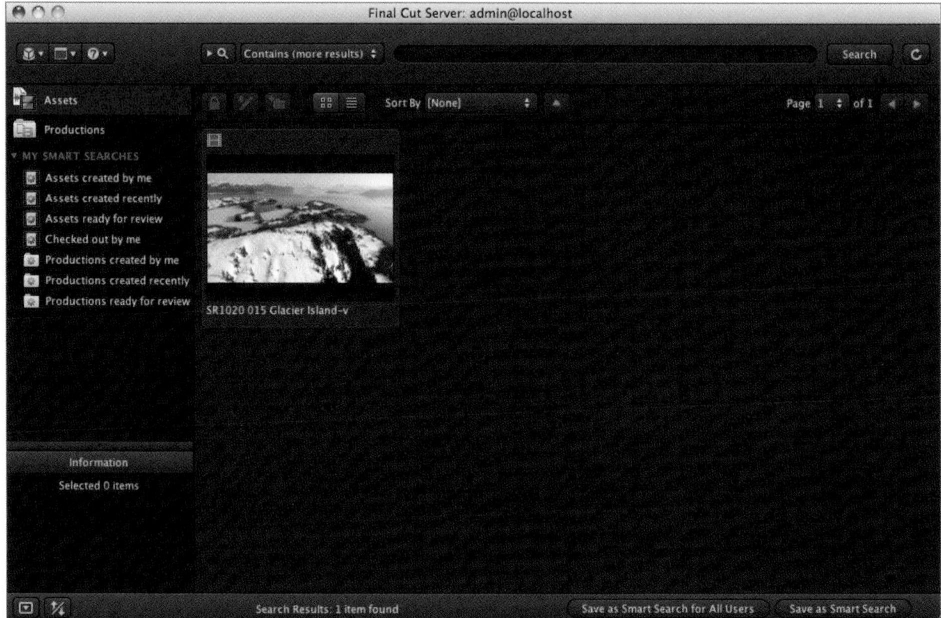

Uploading Using Drag and Drop

If you like to use your mouse, using drag and drop couldn't be easier. Simply drag a file, a series of files, or an entire folder to the client application window.

1 Using the Finder, navigate to the FCS_Book_Files > For Upload folder on the desktop.

2 In the For Upload folder, select the **SR1011 050 Copper River Delta-v** and **SR1015 002 Copper River Delta-v** files.

3 Drag both files into the Assets pane of Final Cut Server.

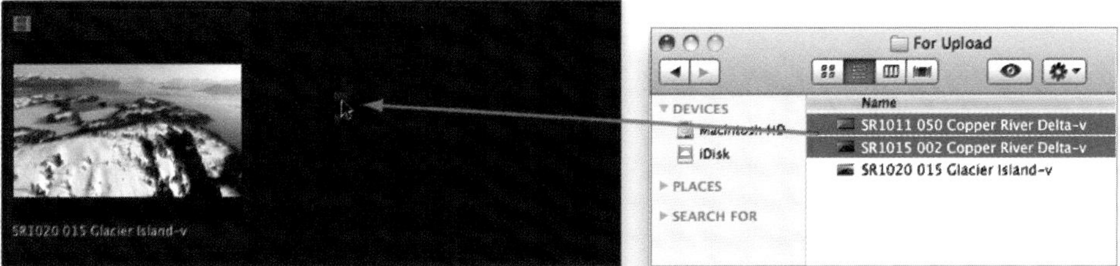

The Upload window appears again; however, this window is titled Multiple Upload. The Filename and Title metadata fields are not visible; Final Cut Server will use the existing filenames of each file. The rest of the window functions as the standard Upload window used before. Any metadata you enter will be applied to both assets.

4 Ensure that Destination is set to Media and that Metadata Set is set to Media.

5 For Description, type *Birds flying at the Copper River Delta*. Press Control-Tab.

6 For Keywords, type *Aerial Sea Birds*. Press Control-Tab.

7 Enter your name in the Owner field.

8 From the Status lookup, choose New.

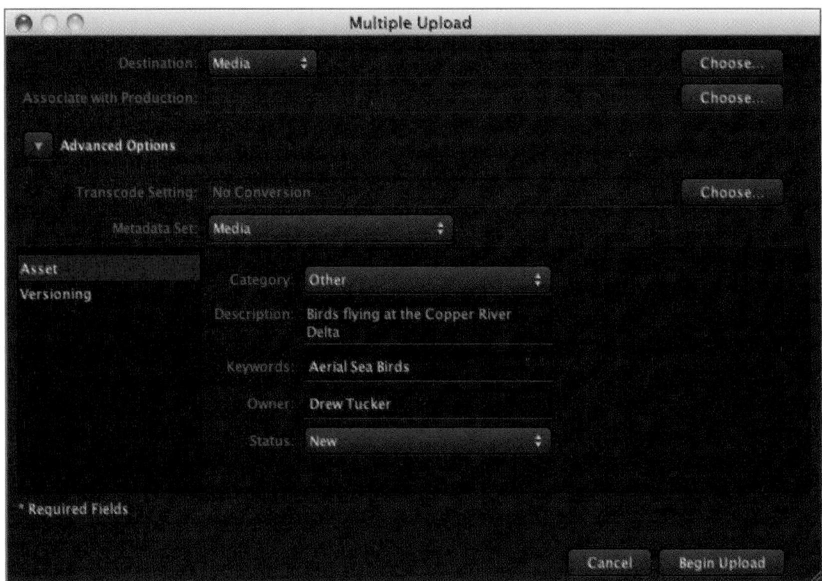

9 Click the Begin Upload button.

This time, the Jobs in Progress notification counts down as each asset and associated representation complete uploading.

10 Click the Refresh button to update the Assets pane.

TIP You can click the Refresh button while the upload is in progress.

Both new assets join the previous asset in the catalog.

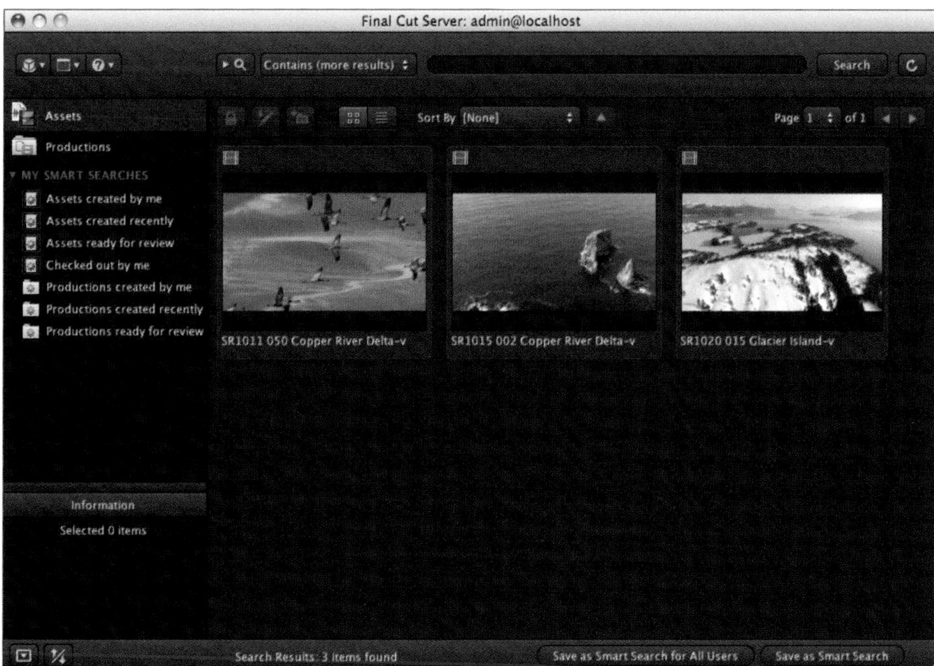

Using Special Uploads

So far you have uploaded self-contained QuickTime movie files. The upload of stand-alone files such as QuickTime movies, music, and documents is straightforward. Files that are dependent on other media files will present other options or require special handling.

NOTE ▶ If you did not copy the FCS_Book_Files folder to the desktop, the following upload may fail.

1 In the Finder, navigate to the FCS_Book_Files > QT Reference folder. Drag the **QT Reference Movie.mov** file to Final Cut Server to upload it.

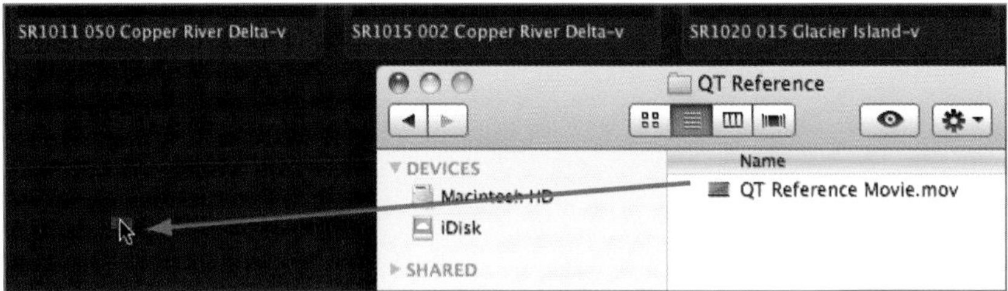

A warning dialog will appear because you are uploading a QuickTime reference movie (which refers to other media content instead of containing all the elements within itself).

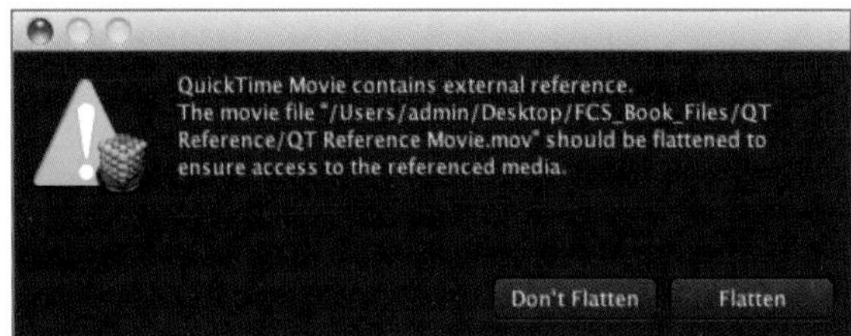

QuickTime reference movies may save storage space because the source media files are referenced rather than copied. If all users who will access this reference movie are connected to your organization's SAN, and the referenced source media is available on that SAN, choose Don't Flatten.

2 Click Don't Flatten to upload the file as a QuickTime reference movie.

3 In the Upload window that appears, click Upload.

Take note of how long this upload takes. After the Jobs in Progress notification has disappeared, proceed with the next step.

4 Drag the **QT Reference Movie.mov** a second time from the Finder to Final Cut Server.

5 Click Flatten.

A progress bar will appear as Final Cut Server creates a self-contained version of the media file.

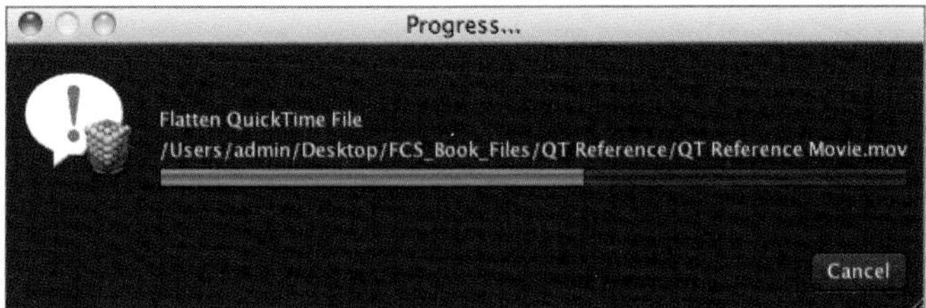

6 When the Upload window appears, change the filename to *QT Reference Movie FLAT.mov*. Click Upload.

Again, take note of how long this upload takes. Depending on your hardware, this upload takes about twice as long, plus the time required to flatten the file.

7 Click the Refresh button to update the Assets pane.

Final Cut Server suggests that you flatten reference files before uploading, which creates self-contained versions of the files. Depending on your hardware and network setup, having Final Cut Server flatten a file may take longer than creating a self-contained file before uploading. If you wish to flatten before upload, see "Exporting a QuickTime Movie as Self-Contained" in the *Apple Pro Training Series: Final Cut Pro 7 Quick-Reference Guide*.

Uploading a Folder of Files

If you upload a folder, a dialog appears asking how you'd like to upload the files in the folder. For example, if you're uploading a folder full of images that you would like to individually catalog, you'll want to create individual assets. However, if you've collected files into a folder for a specific purpose, you may wish to create a single bundle asset for all of them.

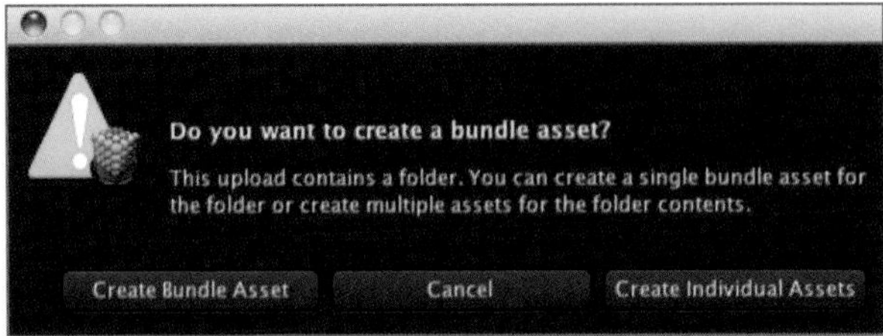

NOTE ▶ The individual components of a bundled asset cannot be manipulated within the catalog. Final Cut Server simply stores the bundle as one asset.

If the folder has an image sequence inside it, Final Cut Server will automatically add it to the catalog as an image sequence.

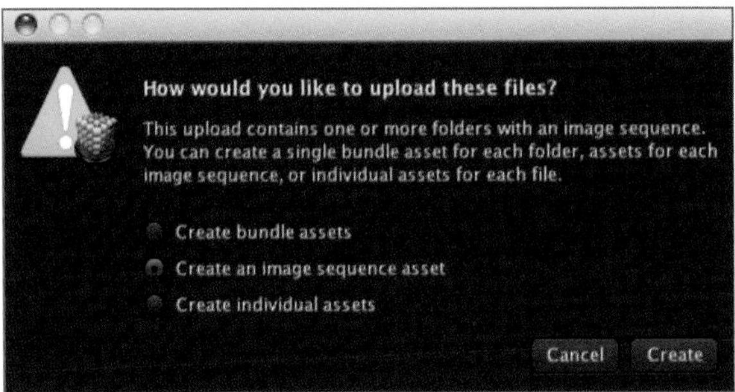

In later lessons, you will deal with Final Cut Studio–specific files (such as Final Cut Pro and Motion project files).

Lesson Review

1. What is the largest metadata container in Final Cut Server?
2. When you upload a folder to Final Cut Server, what are your choices for bundling assets?

3. What are the two ways of manually uploading an asset to Final Cut Server?
4. True or false: When uploading multiple files at once, Final Cut Server allows you to modify the filename of each file.

Answers

1. A metadata set.
2. In the "Do you want to create a bundle asset?" dialog, you can choose to upload the bundle as a single asset, or upload and create individual assets for each of the files within that folder.
3. You may use the Upload File option in the Server pop-up menu or drag a file into the Assets pane.
4. False. The Multiple Upload window will use the existing filenames and does not allow modification.

3

Goals

Create a device with an add-only scan

Modify a metadata set

Automate creation of media assets

Upload a Final Cut Pro project

View metadata of an asset

Lesson **3**

Creating Assets via Final Cut Pro

In a typical post-production workflow, ingest is the first step. You need to acquire your files before you can have Final Cut Server discover and manage them. This lesson takes you through the process of acquiring your assets using Final Cut Pro and having Final Cut Server catalog those assets.

There are two main methods for cataloging FCP assets using Final Cut Server: scan and upload. Both of them have advantages and disadvantages, but choosing between them really comes down to what works best in your workflow.

Before you start ingesting media and creating assets, you need to configure some features of Final Cut Server. Then, you will set up the Final Cut Pro capture scratch folder as an auto-scanned device in Final Cut Server.

Acquisition Scenario #1: Scan

The first scenario involves having Final Cut Server scan the FCP Capture Scratch folder and automatically add assets to Final Cut Server as they are ingested using log and transfer. This has the advantage of making your media available immediately to everyone in your workgroup, and not waiting for the editor to save the project and upload into Final Cut Server. This scenario is used most often in a multiple-user environment where the capture scratch usually resides on shared storage (Xsan). Any metadata entered during the log-and-transfer process is automatically mapped across to Final Cut Server as log and transfer embeds the logging notes into the QuickTime file. Mapping metadata literally means taking external metadata (in this case, Final Cut Pro metadata) and moving it into Final Cut Server.

ADMIN ▼

Preparing Final Cut Server

Before you ingest your assets using Final Cut Pro, you may first set up Final Cut Server to automatically discover these assets. You'll be creating and adding the Final Cut Pro capture scratch location as a device in Final Cut Server, and you'll be setting up add-only scans to pick up this new material for the first scenario.

Final Cut Server ships with default templates for metadata that should work for most people, but also allows you to create your own fields, groups, and sets to suit your needs. You'll be modifying the default asset metadata set (Media) to add the FCP Log and Transfer and P2 metadata groups so that any metadata entered during log and transfer will automatically be mapped to Final Cut Server.

1 From the Apple menu, choose System Preferences, and then select the Final Cut Server pane.

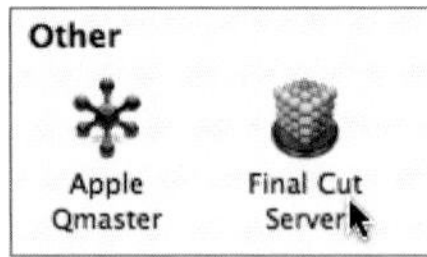

2 In the Final Cut Server pane, click the lock in the lower-left corner. If you're working through this lesson in a classroom setting, authenticate using the credentials provided by your instructor.

3 Click the Devices tab, and click the Add (+) button in the lower-left corner to add a new device.

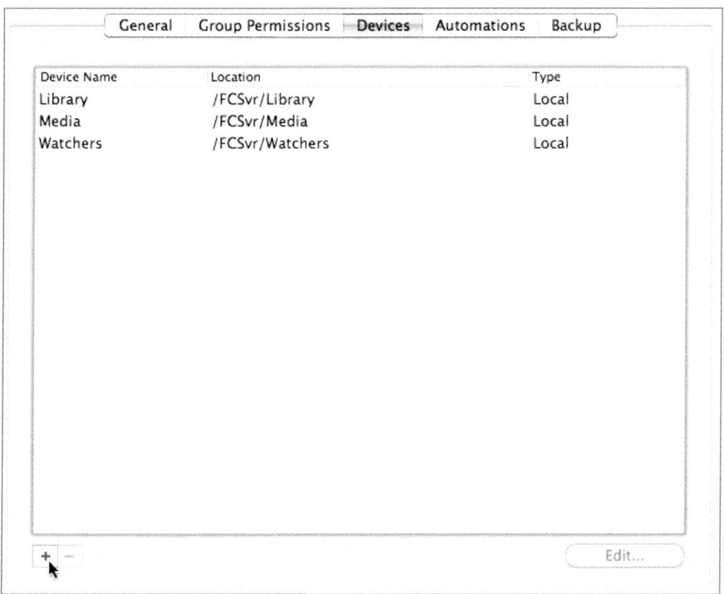

The Device Setup Assistant opens.

4 Select Local for device type and click Continue.

NOTE ▶ The three types of devices are local, network, and Xsan. A local device would be a folder that resides on the server itself, a network device would be a supported (AFP, NFS, FTP, or Samba) network file system, and an Xsan device would be a folder that resides on an Xsan (the Apple SAN solution) file system.

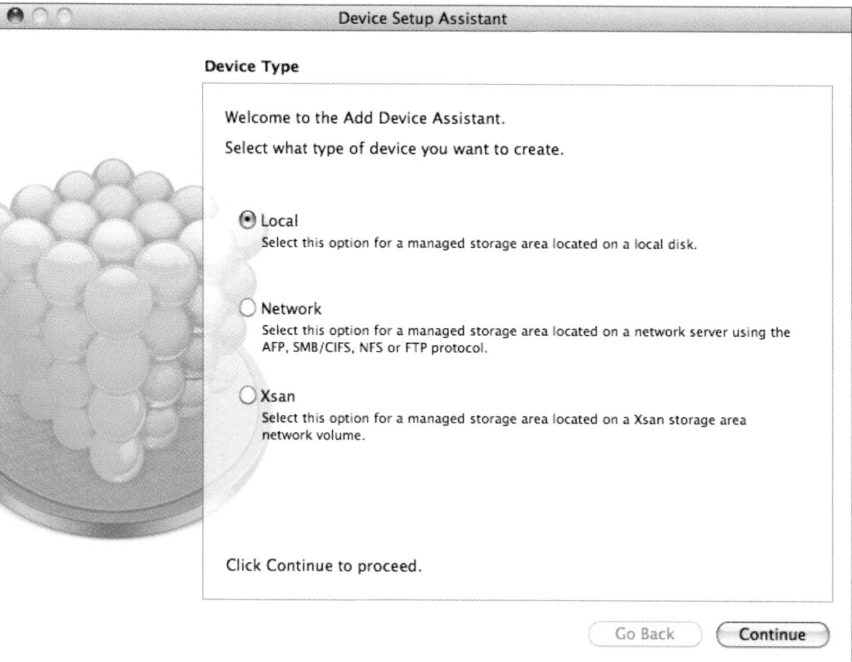

5 Enter *FCP Scratch Disk* as the device name.

6 For Location, click Browse. Navigate to Macintosh HD and create a new folder named *FCP Scratch Disk*. With the new folder selected, click New Folder again and name the folder *Capture Scratch*. Click Choose.

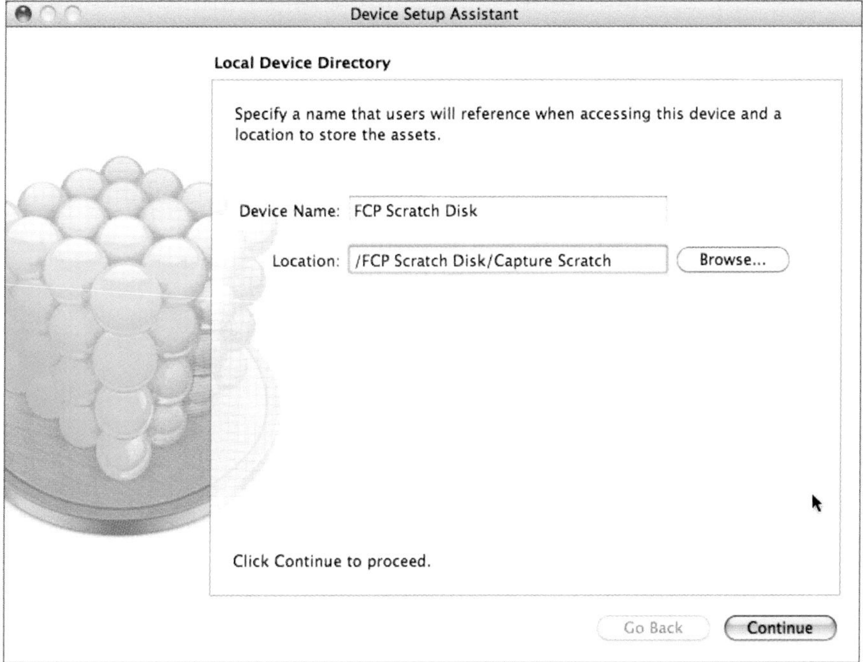

7 Click Continue.

8 Leave "Enable as an Archive Device" deselected and click Continue.

NOTE ▶ An archive device is a special type of device that is flagged by Final Cut Server to allow for the archiving of assets. You'll learn more about creating archive devices in a later lesson.

The Scan Settings page that appears allows you to create a schedule for automatic cataloging of the device. In this exercise, you will create a scan to catalog new media files ingested to this device by Final Cut Pro.

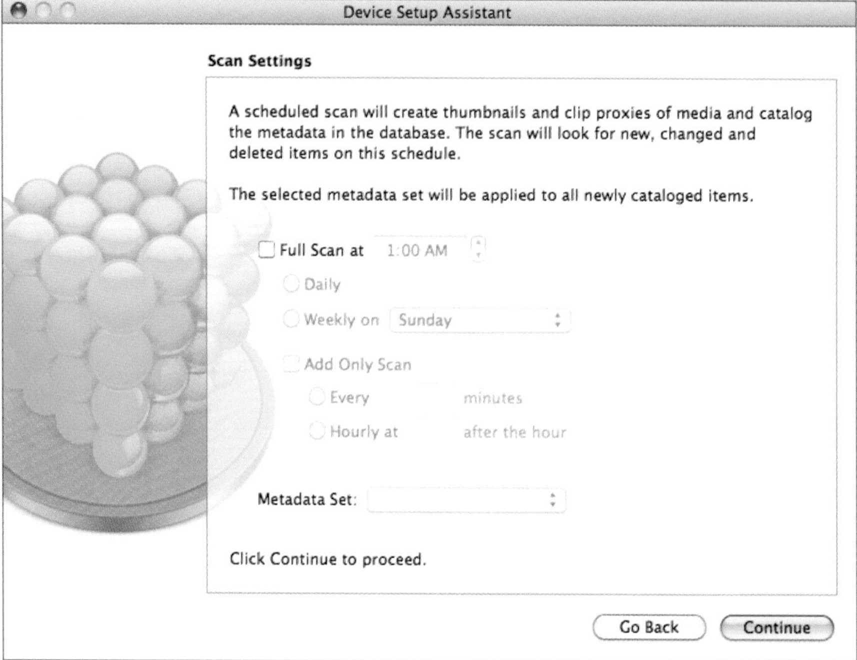

9 Select the Full Scan checkbox and leave the time at the default.

A full scan not only adds new assets, but it also deletes assets from the catalog if they are deleted from this device. Running full scans more than once a day is not recommended, as they're very file system/database-intensive and can affect overall Final Cut Server performance.

10 Select the Add Only Scan checkbox, and change the interval to 5 minutes. Leave Metadata Set as Media (the default), and then click Continue.

NOTE ► Setting the Add Only Scan value to a short interval will not have a significant impact on performance.

NOTE ► Transcode settings in Final Cut Server define what kinds of formats and codecs you'd want to convert files into when you place them on this device. You can also use the transcode settings to specify the format in which files are uploaded and downloaded.

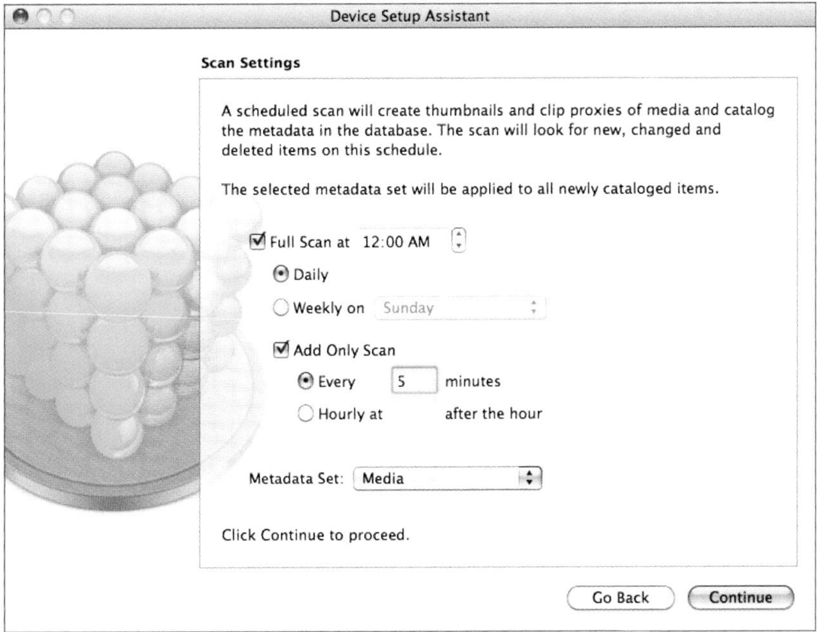

11 You won't be publishing any media to this new device using Final Cut Server, so you can leave the default transcode settings as No Conversion. Click Continue.

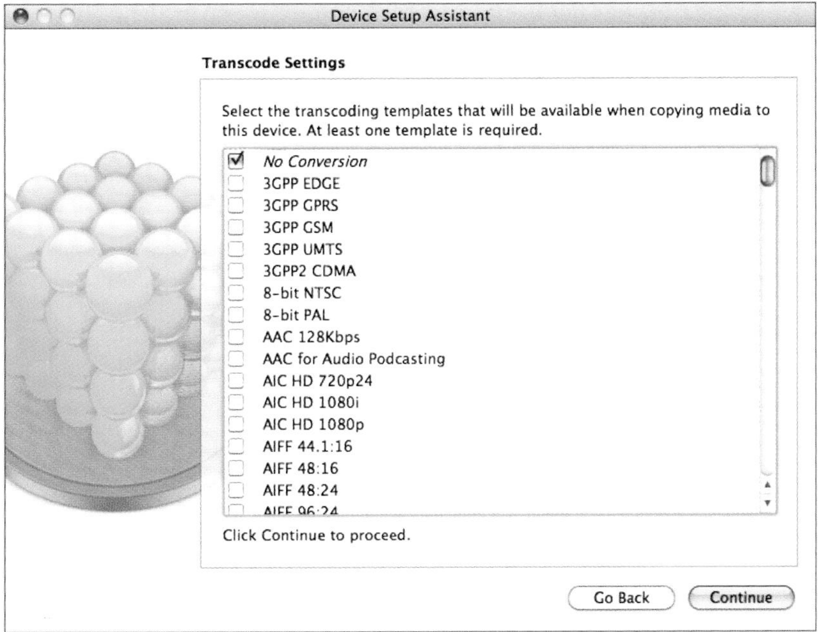

12 On the summary page, click Done and quit System Preferences.

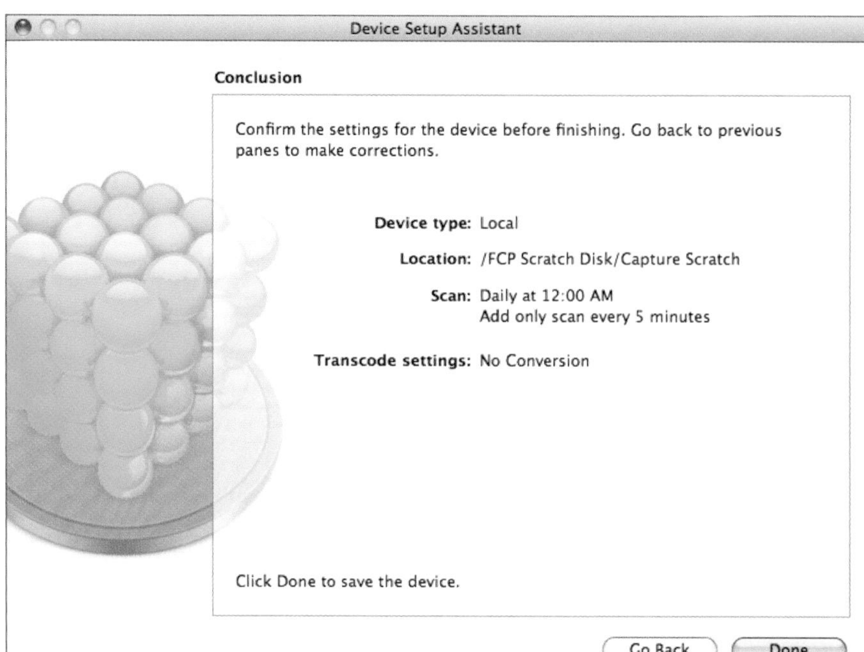

Final Cut Server will now scan this device every 5 minutes for new media files and create assets for those files. Now you will configure Final Cut Server to use metadata created inside the Final Cut Pro Log and Transfer window to follow the new files as they become assets in the catalog.

13 Double-click the Final Cut Server icon on your desktop to open the client application and log in as administrator.

14 From the Server pop-up, choose Administration.

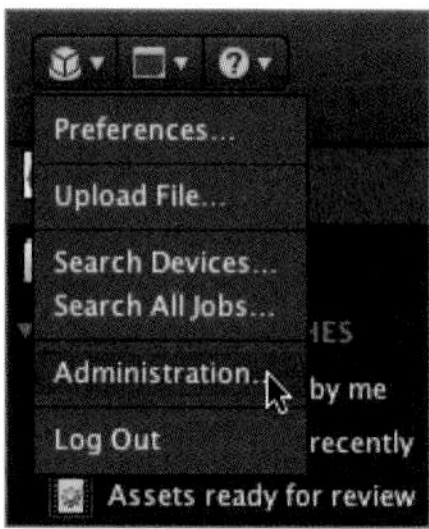

A warning dialog appears cautioning you of the potential hazards of using the Administration panel.

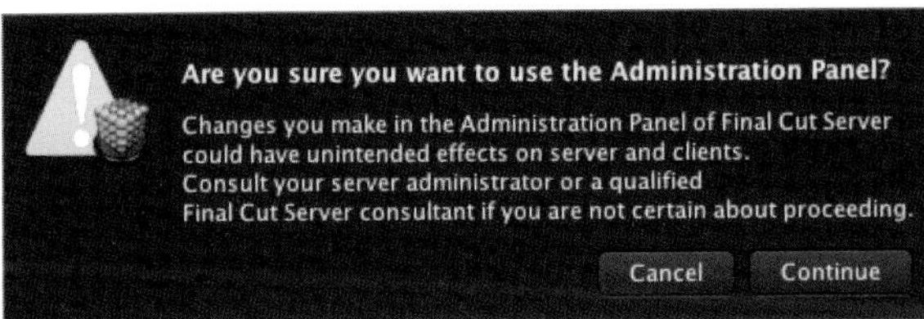

15 Click Continue.

The Administration panel opens.

16 In the Administration panel, choose Metadata Set from the sidebar on the left.

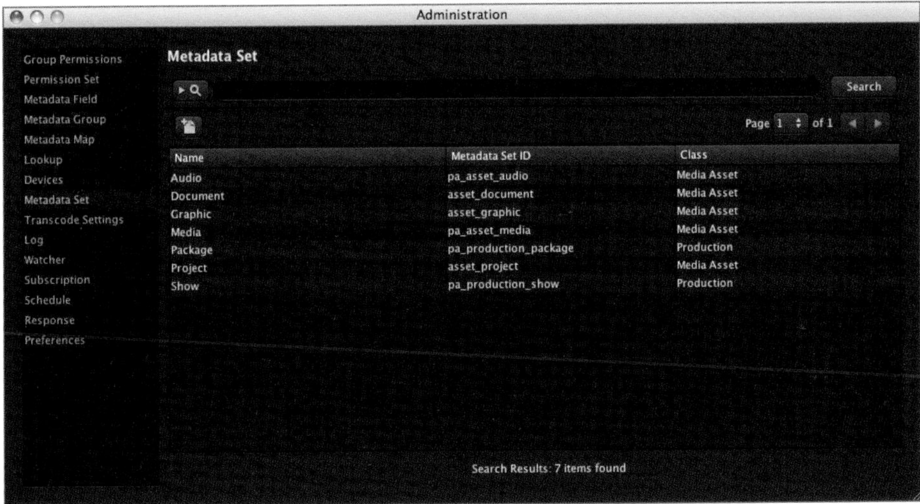

NOTE ▶ The Metadata Set pane is not accessible from System Preferences.

17 Double-click the Media metadata set. Choose FCP Log and Transfer from the Available list on the right and click Add.

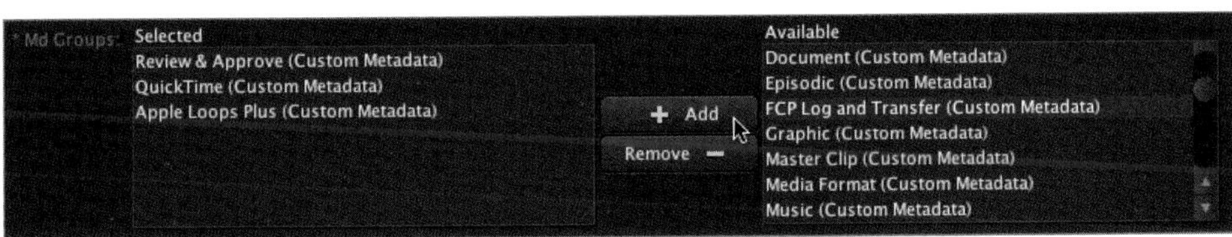

By doing this, you're modifying the default Media metadata template and making the Final Cut Pro Log and Transfer metadata available to Final Cut Server.

The source footage you will ingest in Final Cut Pro comes from a Panasonic P2 card. You can make the Panasonic P2 metadata found in Final Cut Pro available to Final Cut Server.

18 Choose P2 and click Add.

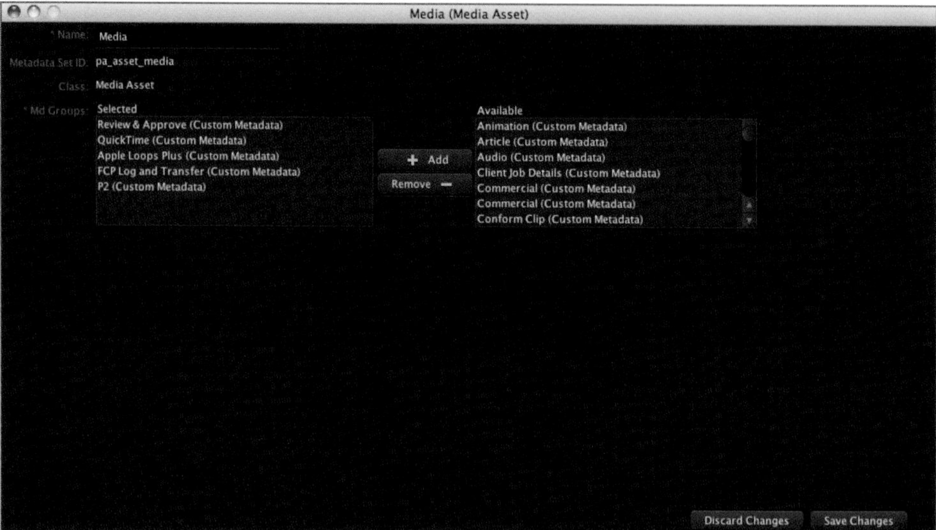

Both items now appear in the Selected list on the left.

19 Click Save Changes. Close the Administration window and log out of Final Cut Server.

You're now ready to begin ingesting assets and have the Final Cut Pro Log and Transfer and P2 metadata automatically mapped to Final Cut Server.

USER ▼

Assigning the Scratch Disk to a Scanned Device

Before you ingest clips into Final Cut Pro, you'll need to configure some settings for Final Cut Pro. You'll set the capture scratch location for Final Cut Pro to the device you created in the earlier exercise.

1 Open Final Cut Pro from either your Dock or your Applications folder.

If this is the first time you have launched Final Cut Pro, there are a few settings to choose. To simplify this process, you'll accept the initial defaults and then change the capture scratch location.

2 In the Choose Setup dialog, leave everything at default and click OK.

NOTE ► If the Choose Setup dialog did not appear, you have previously used Final Cut Pro. Continue to step 3.

3 If the External A/V warning dialog appears, select "Do not warn again" and click Continue.

NOTE ► If you receive a Non-Writable Scratch Disks warning, click Reset Scratch Disks, click OK, and then skip to step 6.

4 From the Final Cut Pro main menu, choose System Settings.

5 In the System Settings dialog, for scratch disks, click the Set button at the top of the list. Navigate to Macintosh HD and select the FCP Scratch Disk folder you created earlier. Do not select the Capture Scratch folder inside.

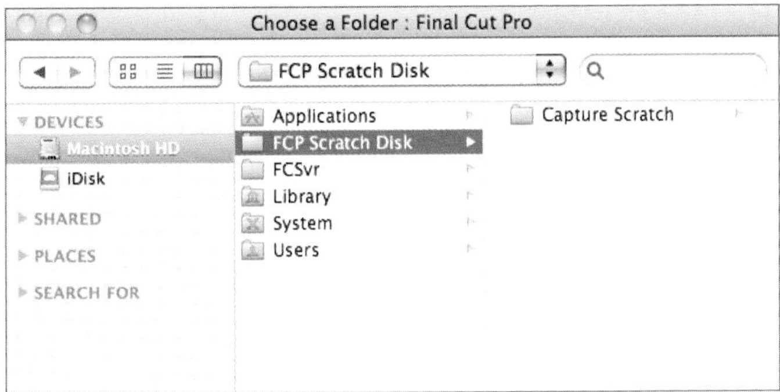

NOTE ► Final Cut Pro will automatically direct transferred files to the Capture Scratch folder inside the selected FCP Scratch Disk folder. If you select the Capture Scratch folder, FCP will create a new Capture Scratch inside the existing Capture Scratch folder.

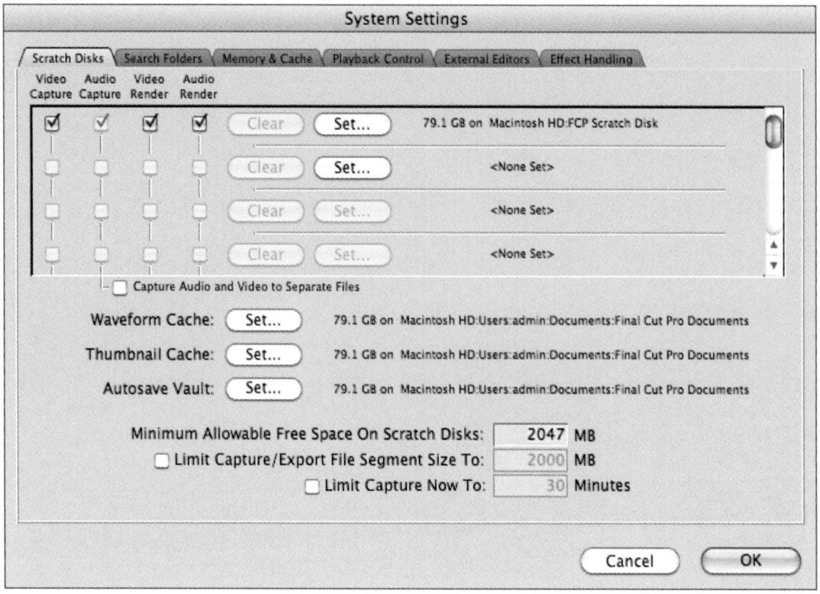

6 Click OK.

Final Cut Pro is now configured to ingest newly acquired media to the location set up in the previous exercise.

> **TIP** Leave Waveform Cache, Thumbnail Cache, and the Autosave Vault at their default, local locations. These folders should always remain on the local system and not be assigned to any network location.

USER ▼

Using Log and Transfer with a Scanned Device

In this exercise, you'll be using log and transfer to acquire some P2 footage, and will have Final Cut Server automatically catalog the assets from the capture scratch location. You won't be moving any of the assets from their primary locations; they'll remain in the capture scratch device. This approach is especially useful in shared storage environments.

1 Using the Finder, navigate to the FCS_Book_Files folder on your desktop. Open **Panasonic P2 Ducati.dmg**. This mounts the disk image file on your local system.

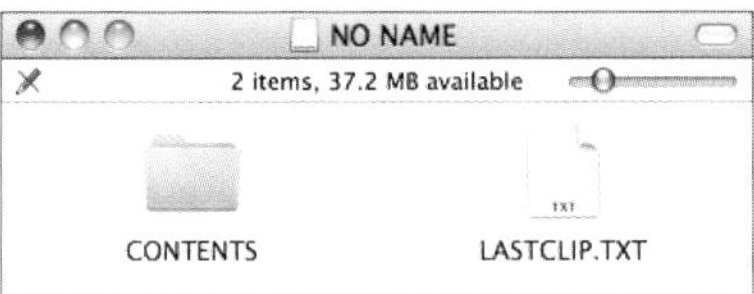

2 Go back to Final Cut Pro, choose File > Save Project As, and save the project to your desktop as *Ducati Project.*

TIP To access the Save command, the Viewer window must not be the active window.

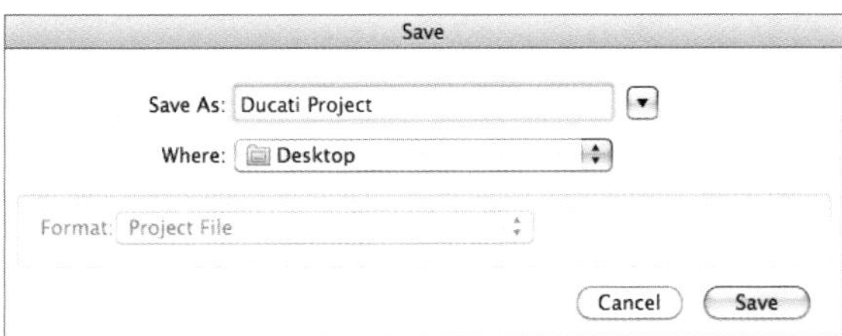

3 In Final Cut Pro, choose File > Log and Transfer (or use the keyboard shortcut Command-Shift-8). If you don't see any media in the Log and Transfer window, verify that you have mounted **Panasonic P2 Ducati.dmg** correctly.

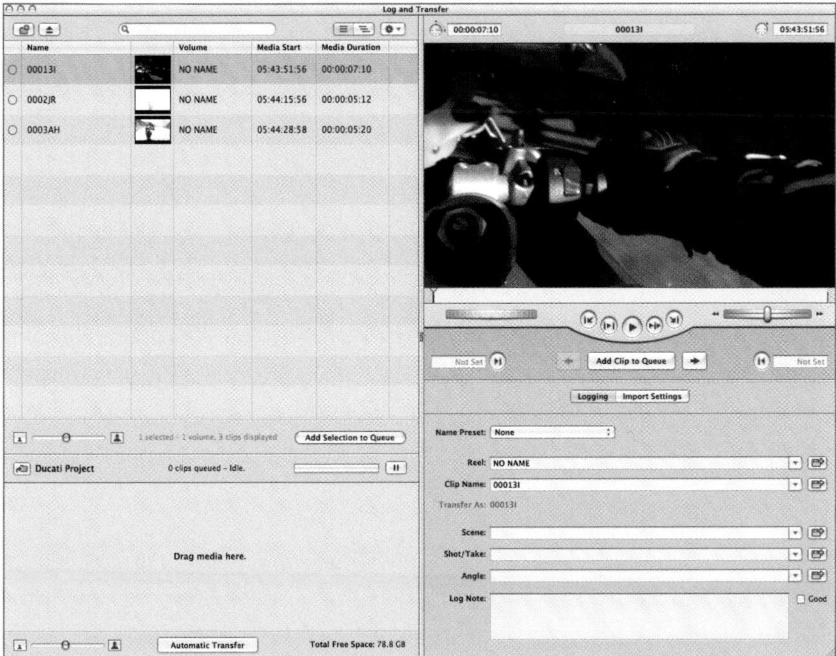

4 If necessary, select the first clip in the list (**0013I**), which loads the preview in the Viewer at the top right of the Log and Transfer window.

5 At this point, you can enter some metadata in the Logging tab in Log and Transfer. For Reel, enter *Ducati Project*. For Clip Name, enter *ducati1*. For Scene, enter *7*; for Shot/Take, *2*; for Angle, *A*; and for Log Note, enter *Initial Ducati clip*. Finally, select the Good checkbox to the right of the Log Note field.

Typing in this metadata allows you to find your clips more quickly in Final Cut Server, as you will have more search terms.

NOTE ▸ You can enter your own values for most of the Log and Transfer metadata fields, but remember that Clip Name will be the filename that is given to the media being logged and transferred. Also, remember that any metadata you enter into Log and Transfer will be automatically embedded into the QuickTime file and thus picked up during discovery by Final Cut Server. This will prevent users from having to enter metadata more than once and will assist them in locating the media inside Final Cut Server.

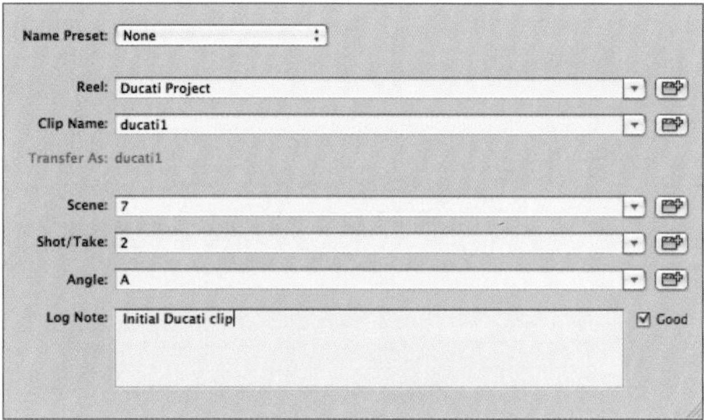

6 Click the Add Clip to Queue button, and note that Final Cut Pro places this clip into a queue that will transfer its media to a QuickTime file in your scratch disk location.

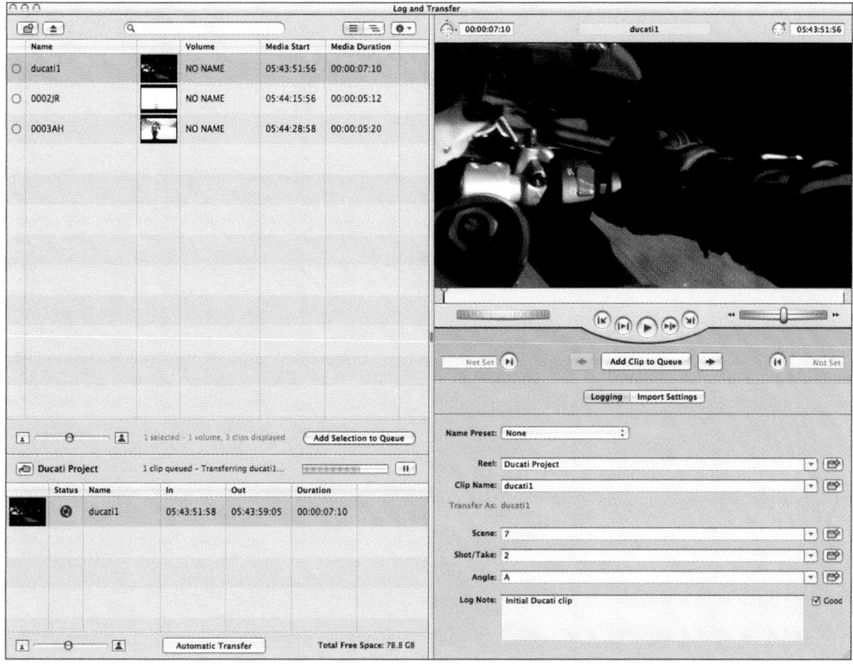

7 After the transfer has completed, close the Log and Transfer window, and then quit Final Cut Pro. Click Yes to save the project.

8 In the Finder, open your FCP Scratch Disk directory and you'll notice that inside Capture Scratch, you now have a Ducati Project folder, and inside that you have your acquired Ducati movie (**ducati1.mov**).

After five minutes, the Add Only Scan will catalog the new media found in the capture scratch you set up earlier as a device.

9 Open the Final Cut Server client application, and log in.

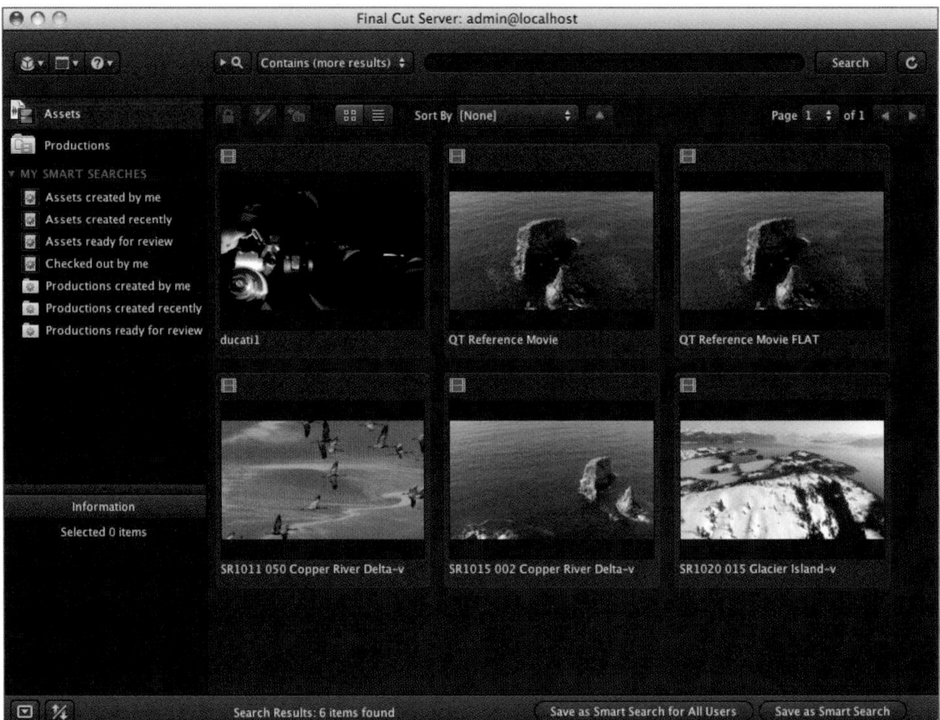

You will find the **ducati1** thumbnail as the first asset in the catalog (the sort defaults to alphabetical). Final Cut Server has automatically cataloged the media ingested through log and transfer. This allows others in your workgroup to begin working with the content right away, instead of having to wait for the editor to finish the ingest process.

TIP If the asset has not yet appeared, wait a minute and click Refresh.

Though the catalog is fairly small, you can use a simple search to zero in on the Ducati clip.

10 In the search field, type *ducati* and press Return.

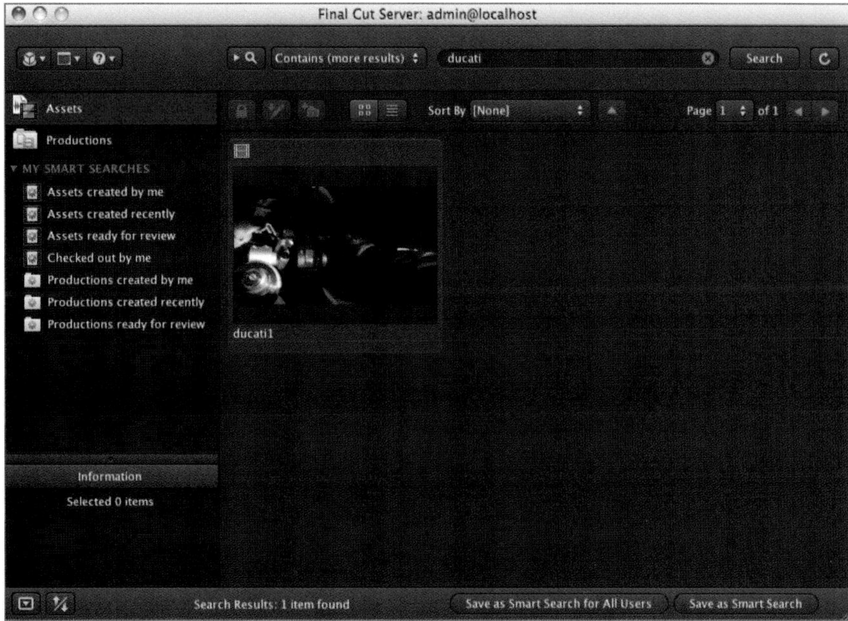

11 Double-click the **ducati1** asset to inspect the attached metadata.

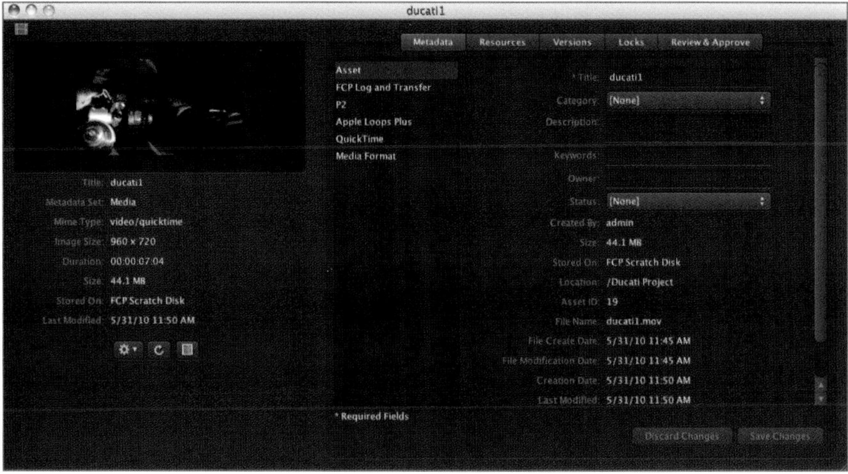

The asset's information window opens. Here you can find an asset's details such as the metadata associated with the file (file size, where it is stored, and so on) and any metadata you have assigned to the asset.

12 In the list on the left of the Metadata pane, click FCP Log and Transfer. Notice that the metadata you entered during log and transfer has been automatically transferred from Final Cut Pro to the asset in Final Cut Server.

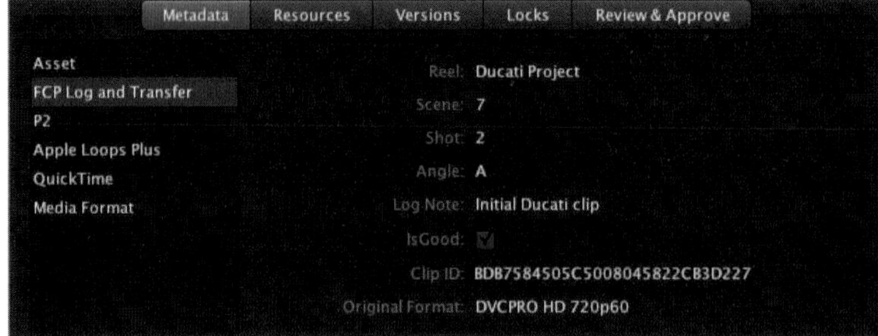

TIP Entering the metadata during ingest will save your organization time in the long run, as you will not have to enter metadata more than once. Entering the proper amount of metadata ensures that your assets will be easy to locate and work with.

13 In the list on the left of the Metadata pane, click P2. Notice that the metadata from the Panasonic camera is transferred because you customized the Final Cut Server metadata template to include Panasonic P2 metadata.

14 Close the ducati1 asset window.

Acquisition Scenario #2: Upload

The second scenario involves using a straight Final Cut Pro upload process through Final Cut Server. In this scenario, Final Cut Server does not scan the capture scratch device, but instead waits for the editors to finish ingesting and logging all of their clips, save their project, and then upload it to Final Cut Server. This copies the files from the local Capture Scratch folder to whatever destination device was specified during the upload process. If the original files resided on an edit-in-place device, they would not be copied, but instead added to the catalog in their current location. This approach has the advantage of tracking all the media together with a project file, but relies on the editor to upload the project. Otherwise, the media won't be seen in Final Cut Server.

The main advantage you gain with this scenario is the ability to manage your ingested material as a Final Cut Pro project file. It also gives you the ability to maintain version control over your Final Cut Pro projects, to enable the use of edit proxies if you require an offline workflow, and to tag your project file and assets with custom metadata that goes above what is presented in Log and Transfer.

USER ▼

Assigning and Using a Local Scratch Disk

You'll begin by reassigning the Final Cut Pro scratch disk away from the device set to automatically scan. You'll then create a new Final Cut Pro project, transfer some additional clips using log and transfer, and finally upload the project into Final Cut Server to be managed with some custom metadata.

1 Make sure the **Panasonic P2 Ducati.dmg** file is still mounted on your system. (You can verify by checking the Finder.) If you don't see it, double-click the disk image to mount the P2 material again.

2 Open Final Cut Pro.

3 Choose Final Cut Pro > System Settings.

4 Click the top Set button, and navigate to the default Final Cut Pro scratch disk at Macintosh HD/Users/admin/Documents/Final Cut Pro Documents. Click Choose, and then click OK to close System Settings.

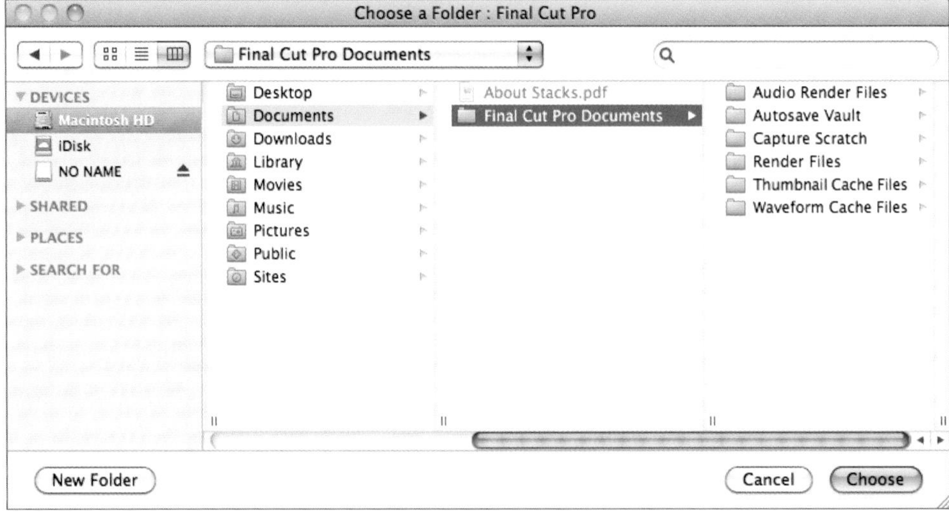

5 Choose File > Close Project. In the dialog that appears, click Yes to save.

6 Choose File > New Project to create a new project.

7 Save the new project to the desktop as *Ducati Managed Project.*

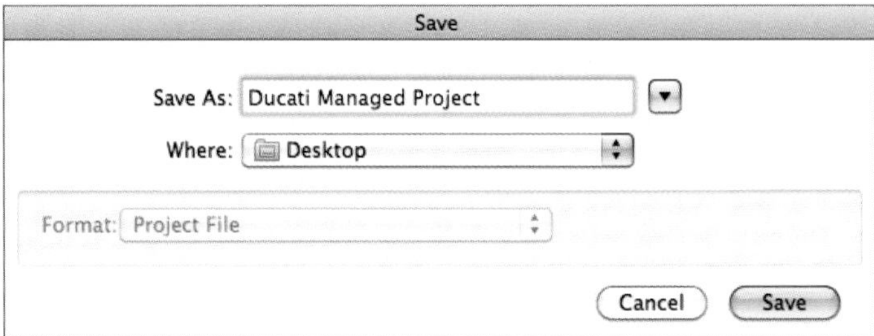

8 From the File menu, choose Log and Transfer.

9 Select the second clip. In the Clip Name field, name the clip *ducati2*, enter additional metadata as shown in the following figure, and then click Add Clip to the Queue.

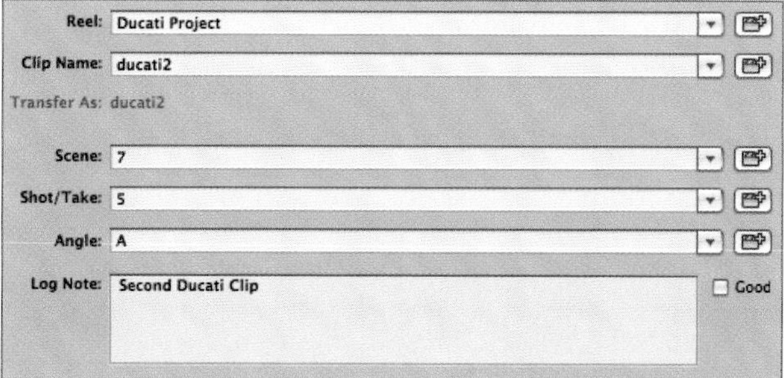

10 Select the third clip. Name the clip *ducati3* and modify the other fields as shown in the following figure, and then click Add Clip to the Queue.

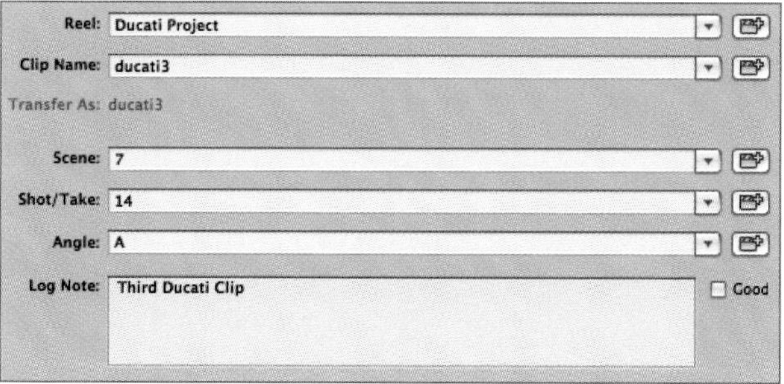

Both clips, **ducati2** and **ducati3**, now reside within the project as shown in the Browser:

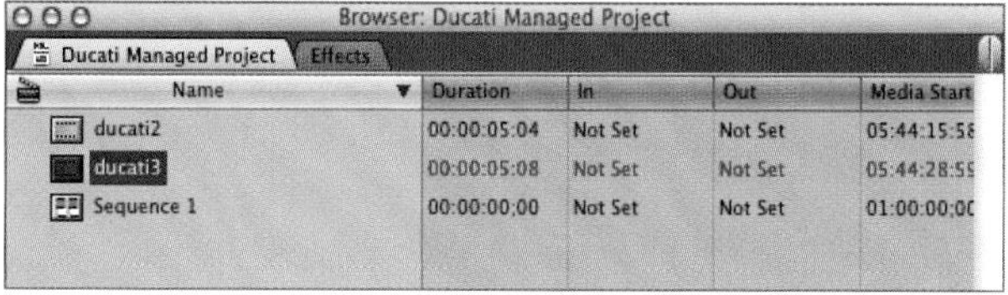

11 Close Log and Transfer. Save your project, and hide Final Cut Pro so that only the Final Cut Server client application window and your desktop are visible.

USER ▼

Uploading a Final Cut Pro Project

You're now ready to upload the Final Cut Pro project file to Final Cut Server.

1 Locate the **Ducati Managed Project** file on your desktop, and drag it anywhere within the Final Cut Server client application window.

The Upload Final Cut Pro Project window appears and prompts you where to save the project and to enter any custom metadata or tags.

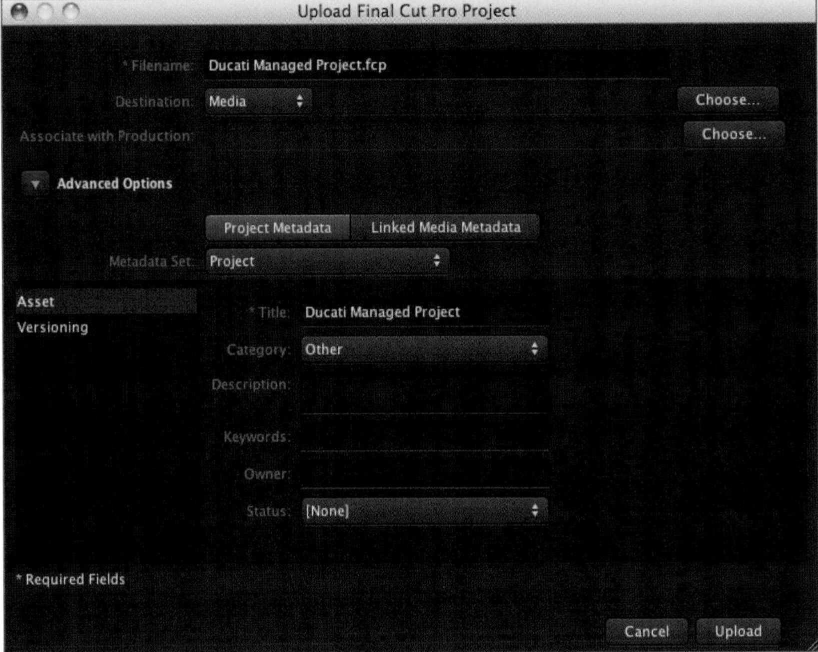

2 From the Destination pop-up menu, choose Media.

3 Ensure that the Project Metadata pane is selected and select Project for the Metadata Set.

4 In the Project Metadata pane, for Description enter *Ducati managed project via upload*; for Keywords enter *motorcycle*; and from the Status pop-up menu choose New.

NOTE ▶ This metadata will be searchable only in the individual project file itself. If you want the tags and metadata to be applied to each asset in the Final Cut Pro project, click the Linked Media Metadata tab and enter the desired information.

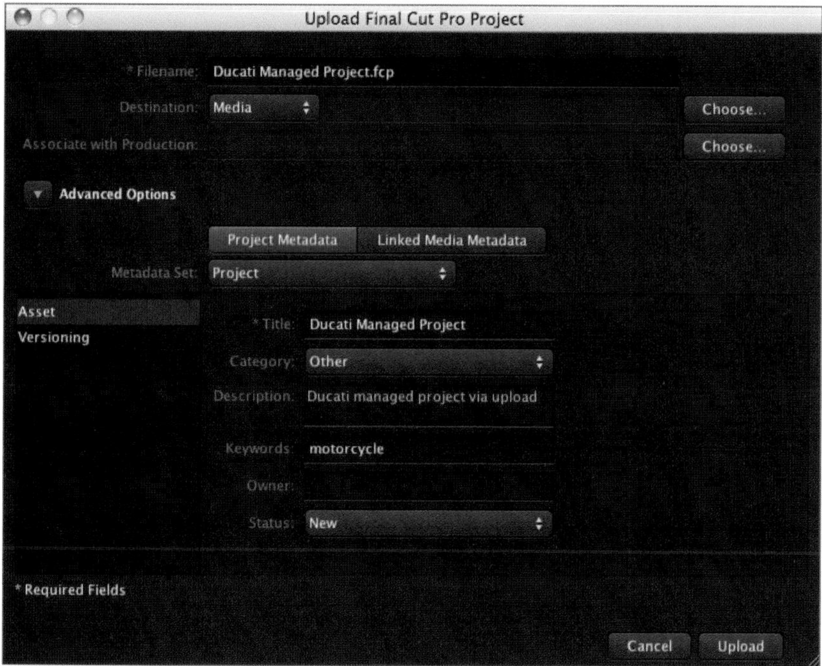

5 Click Upload.

You can monitor the progress of the upload and proxy creation by opening the Search All Jobs window from the Server pop-up menu. During the upload process, edit proxies will be created from your content if you enabled that feature during the installation process, and the clips you ingested will be automatically added to the Final Cut Server catalog.

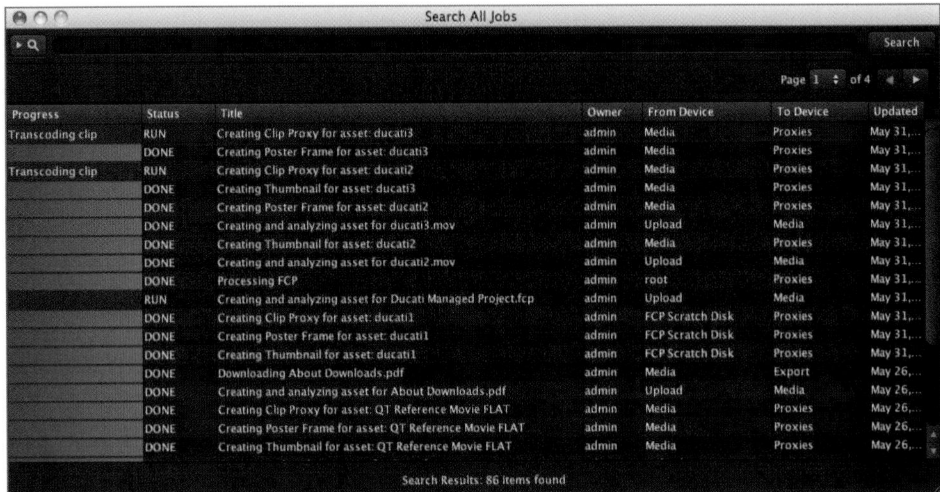

6 After all of your proxies have finished being transcoded, go back to the main search window and click Refresh to update your previous "ducati" search.

Your Final Cut Pro project file now appears and is managed by Final Cut Server.

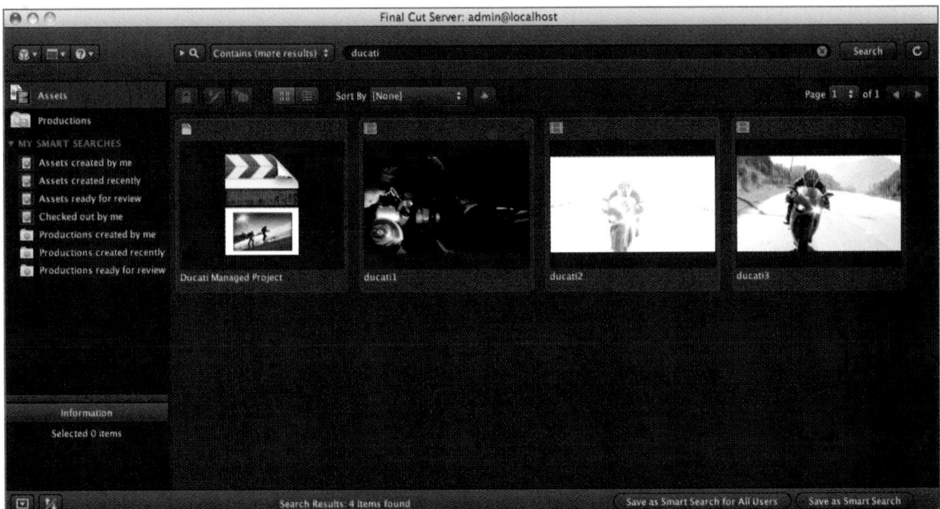

USER ▼

Working with Elements of a Project

During the project file upload, Final Cut Server recognized the new **ducati2** and **ducati3** media files used inside the project. Although **ducati2** and **ducati3** appear as regular assets, they are also linked to the **Ducati Managed Project** file.

1 Double-click the **Ducati Managed Project** asset to open the asset's information window.

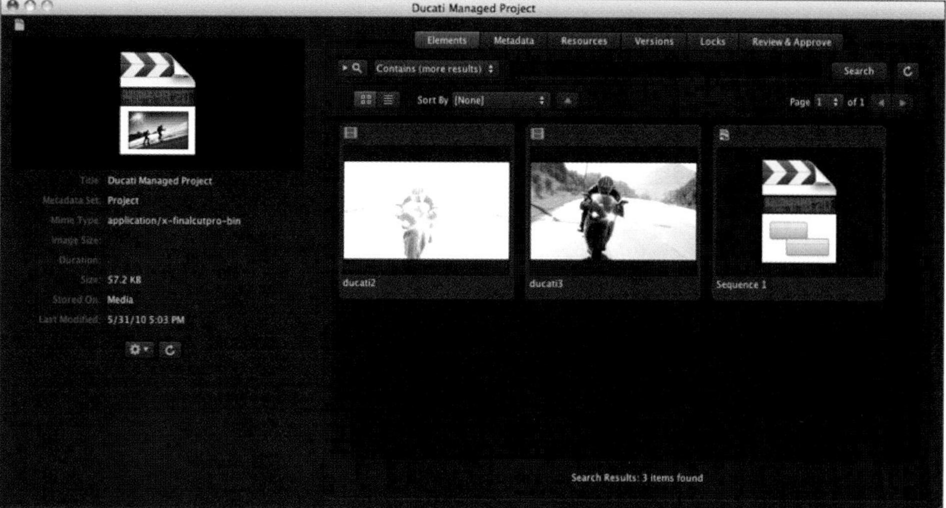

Notice the links to the assets you ingested earlier. In Final Cut Server, these links are called *elements*. Also notice the metadata you entered during the upload process.

2 Click the Metadata tab to see the metadata you entered during upload.

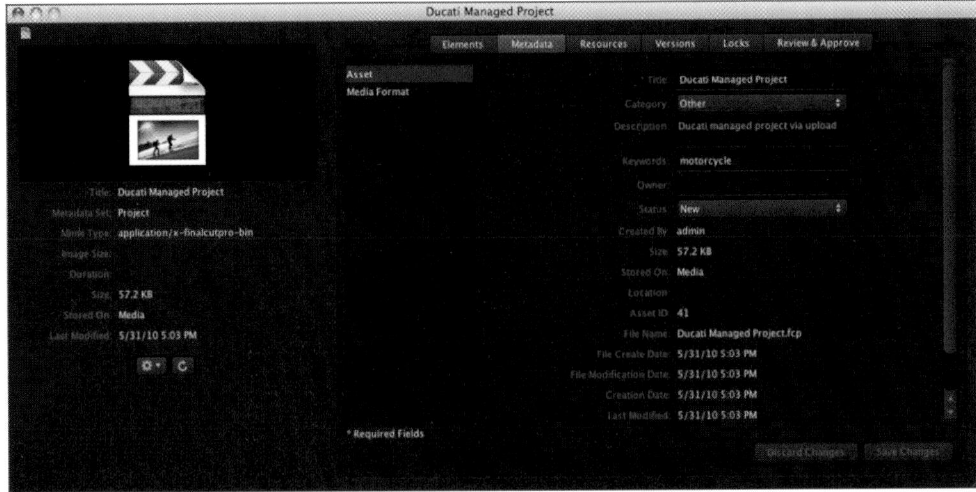

Later in the book, you'll learn about the other features available in this window, including Versions and Review & Approve.

3 Close the asset information window and quit the client application.

Lesson Review

1. What are the benefits of using log and transfer in Final Cut Pro?
2. Which is less database-intensive: a full scan or add-only scan?
3. Where can you view the updated status of a Final Cut Pro project upload?
4. What is the name of the Log and Transfer metadata group?
5. When are edit proxies created?

Answers

1. The metadata entered is embedded directly into the file.
2. An add-only scan is less intensive, as this type of scan creates assets only from files on the device that are not currently cataloged.

3. The Search All Jobs window.
4. FCP Log and Transfer.
5. Only during the upload process of a Final Cut Pro project file.

Goals

- Discover video assets
- Discover graphic assets
- Discover audio assets
- Work with watchers
- Work with scans

Lesson **4**

Creating Assets via Scans and Watchers

In Lesson 3, you learned how to discover and catalog Final Cut Pro assets. This lesson takes you through the process of acquiring other video, graphic, or audio assets that you might have. You'll cover two Final Cut Server automation features: scans and watchers.

Before you dive into the particulars of using the Final Cut Server discovery methods, let's define some key terms:

- A *watcher* is a folder that usually resides on the shared folder that is commonly aliased to a user's desktop and waits for a file to be dragged to it. Dragging a file to the watcher triggers an automated set of responses inside Final Cut Server (usually moving the file dropped in the watcher to another Final Cut Server device).
- A *scan* is a scheduled process whereby Final Cut Server looks at a device or a directory, analyzes the content, and modifies the repository of assets accordingly. If the scan is configured as an add-only scan it will add only newly discovered files to the catalog. If the scan is configured as a purge scan, it removes assets for which the original file no longer exists. If the scan is configured as a full scan, it will do both and also look for modifications in files.

You'll use these terms throughout this lesson as you explain how to access and use the discovery methods.

Using Scans and Watchers

Both scans and watchers can be used to automatically ingest or catalog new assets in Final Cut Server. You have the option of using both or either of them individually in your workflow, depending on the requirements.

You used scans in the previous lesson to automatically add captured Final Cut Pro material to the Final Cut Server catalog. Scans are most frequently used when you have material that already resides on a file system and you wish to leave it in place there. This could be any type of network or direct-attached file system that Mac OS X can read. Most frequently, it will be an Xsan or an AFP/SMB/NFS network that already has an existing file system hierarchy. When you catalog these devices using scans, the folder structure becomes metadata and thus inherently searchable. So scans are good for infrastructure that already exists.

Watchers are most useful when you need to copy and/or transcode content from a location (such as your local edit station) to another location (perhaps an Xsan, or network). Commonly, the watchers are shared on the Final Cut Server and mounted on the client desktops. Local users can then drag material to these folders to be transcoded or published to a final destination. A common workflow would involve dragging a piece of ProRes 422 material to a watcher and having it automatically transcoded to H.264, and then having the resulting H.264 file published to an FTP server or other network device.

ADMIN ▼

Setting Up a Basic Scan

Before you can leverage scans, you must first prepare Final Cut Server for discovering these assets. Final Cut Server can scan only devices, so any area you want to scan has to be defined as a device first. You'll be making a new file-system device to mimic the behavior of a Xsan or network. You will also set up a basic scan to pull in the material from the folder.

1 Copy the Evergreen folder from the FCS_Book_Files folder on your desktop to the Macintosh HD/FCSvr folder.

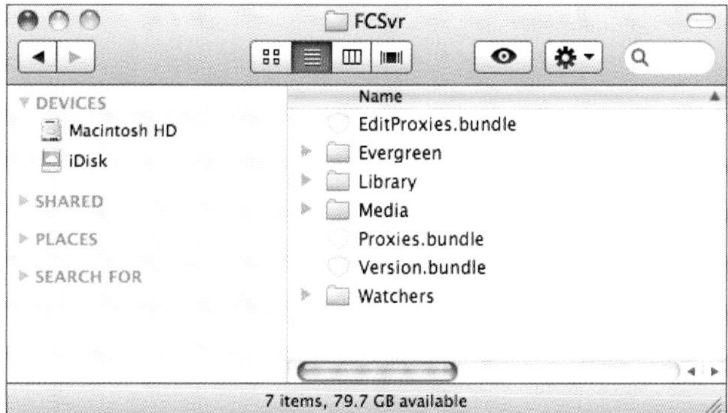

2 Open System Preferences, and click the Final Cut Server icon.

3 Click the lock to authenticate yourself as an administrator and then click the Devices tab.

4 In the Devices pane, click the Add (+) button to add a new device.

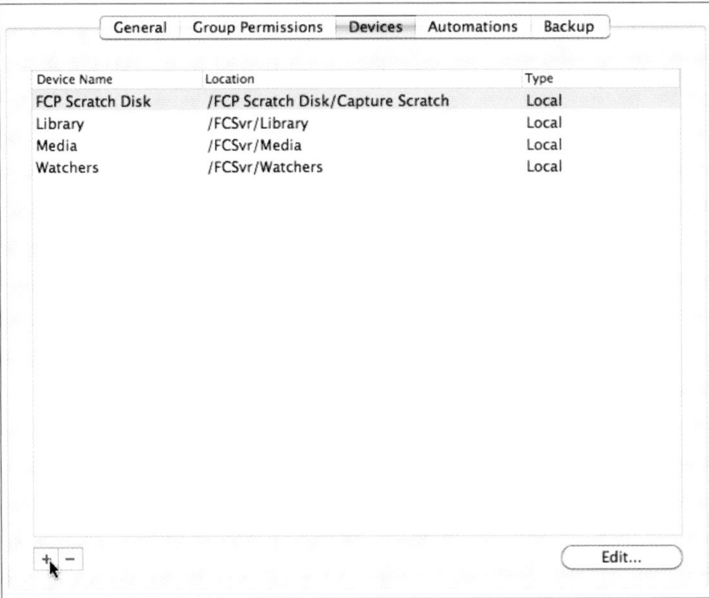

5 In the Device Setup Assistant that opens, select Local and click Continue.

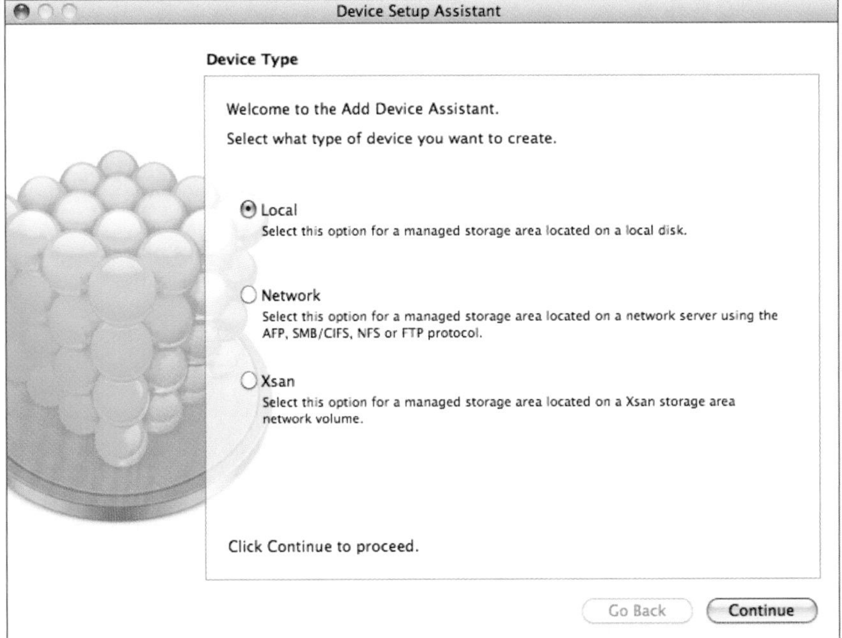

6 Enter *Evergreen* for Device Name, click Browse, and then navigate to the Evergreen folder you copied to Macintosh HD/FCSvr. Click Continue.

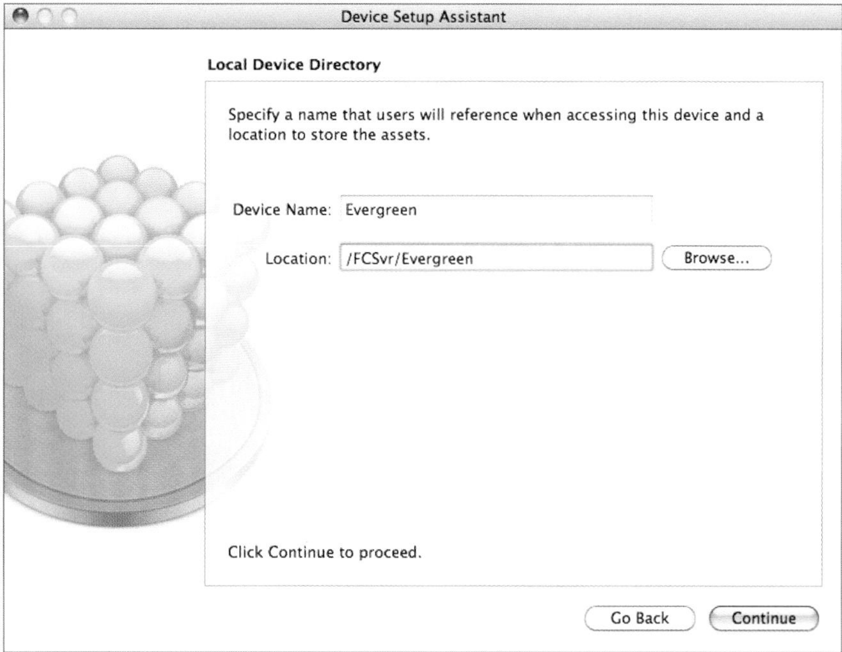

7 In the next dialog, leave "Enable as an Archive Device" deselected and click Continue.

8 In the next dialog, select the Full Scan checkbox and set the kickoff time to be approximately 5 minutes after your server's current time. Select the Add Only Scan checkbox, and set it to kick off every 5 minutes. Leave Metadata Set at Media and click Continue.

NOTE ▶ We've specified a full scan here strictly for the purposes of this example. You should *never* configure a full scan to run during the middle of the day, as it will consume bandwidth and take resources away from other processes. Full scans should be scheduled only during off hours when production is not critical.

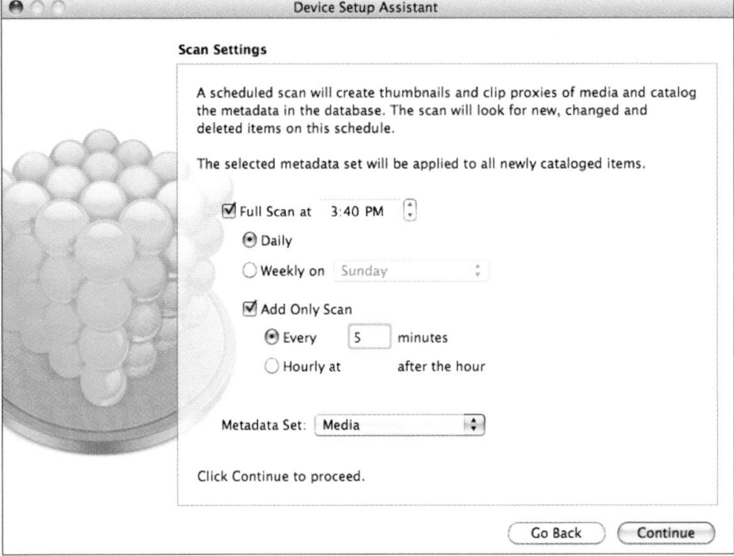

9 In the next dialog, leave No Conversion selected. Because in this exercise you're using this device as a source device, not a destination device (which might require a transcode), you don't need to select any other codecs. Click Continue.

10 Review the final settings, and make sure they're similar to those in the next screen shot (your time will differ). Click Done when you're finished comparing. If you've made any mistakes, you can click Go Back and fix them.

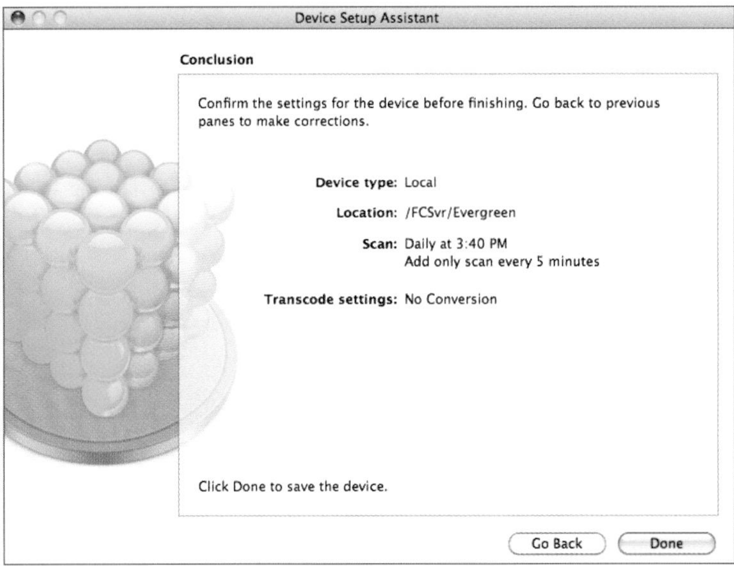

11 Close System Preferences.

12 Open the Final Cut Server client application and log in as an administrator.

Depending on what time you set your scans to run, within 5 minutes Final Cut Server will start automatically discovering assets from the new device and adding them to the catalog.

13 Choose Search All Jobs from the Server pop-up menu to watch the proxies get created in real time.

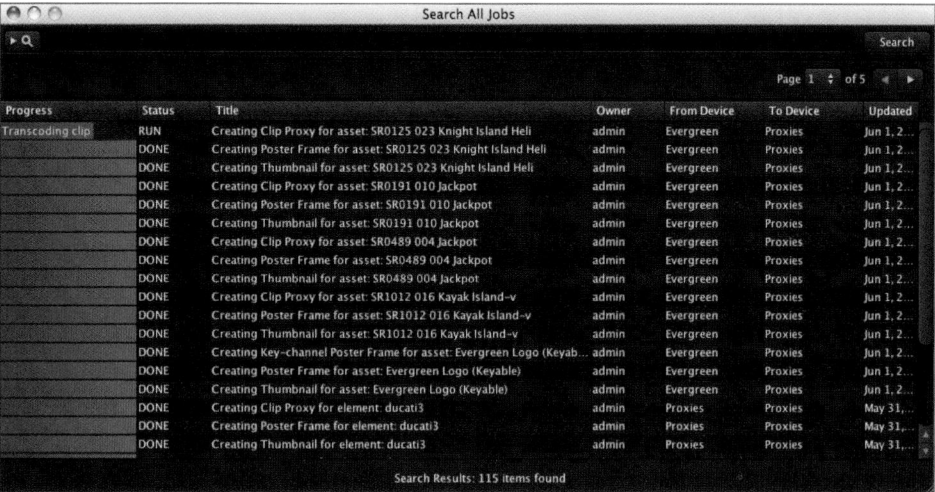

14 Close the Search All Jobs window.

Back in the Assets pane, you'll notice that the new Evergreen assets are not visible. By default, Final Cut Server remembers your last search (even after relaunching the client). You need to clear the last search to reveal the current catalog.

15 At the right side of the search field, click the Clear button.

You'll notice that you've pulled in different types of assets: graphics, audio files, and movies. In an upcoming exercise, you'll set up Final Cut Server to pull in these different types of files tagged based on filters, which will give you greater flexibility when it comes to automated metadata tagging.

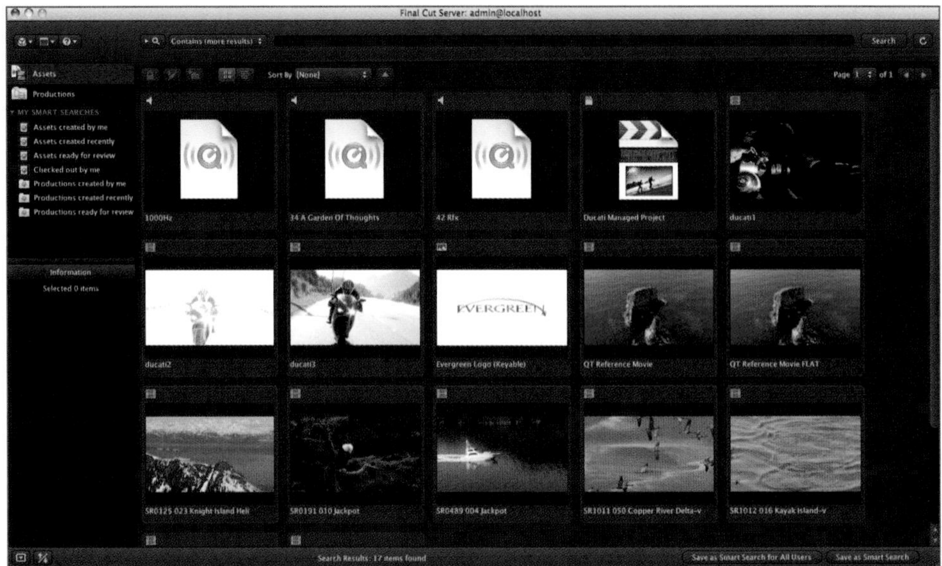

ADMIN ▼

Breaking Down the Scan

The scans you created in the previous exercises/lessons (and in the watchers coming up) are known as automations. An *automation* is a combination of a trigger and a response. The triggers you've used so far were schedule-based (set to happen at a specific time or at a recurring interval). As you'll find out later, triggers can also be based on metadata or user actions. These triggers are used to activate a response. A *response* is an action in Final Cut Server that can do a number of things. The most basic response is the scan, as you configured earlier. You'll see throughout this book that the responses are where the true power of Final Cut Server comes alive.

Triggers and responses are configured in multiple areas of the Administration window of the client application. For now, let's take a look at the scan responses that were created when you added the Evergreen device through System Preferences.

1 Open the Administration window.

2 From the menu on the left side of the window, choose Response, and then double-click "Scan device Evergreen [Full Scan]."

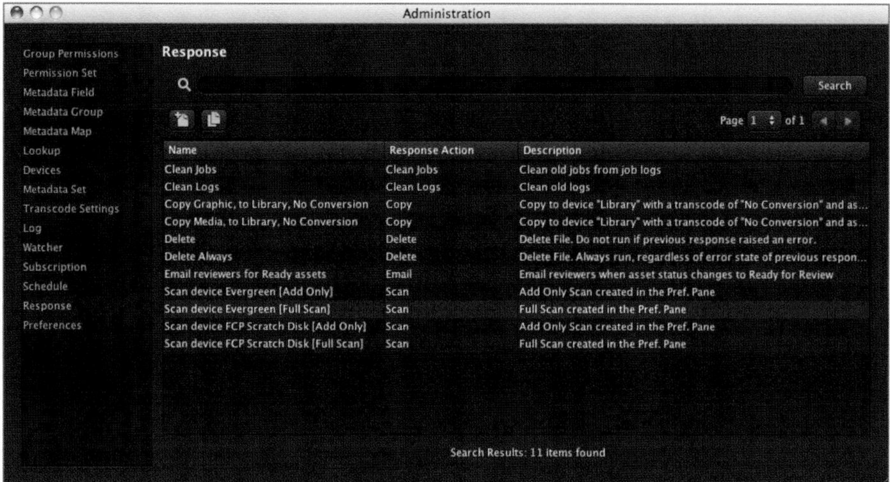

3 On the left side of the window, click Scan and scroll down.

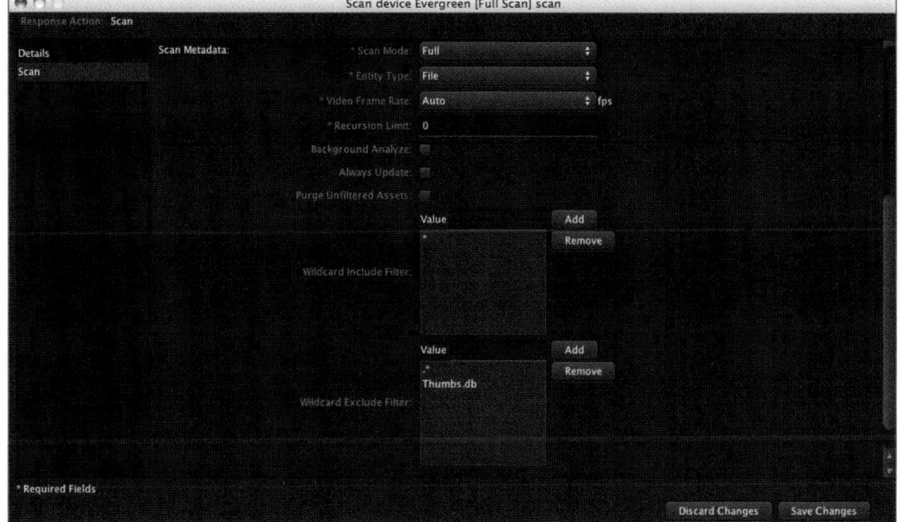

Notice the values that were inserted by System Preferences. Of specific interest are the Wildcard Include Filter and Wildcard Exclude Filter values. The ".*" and "Thumbs.db" in the Wildcard Exclude Filter prevent the scan from inadvertently picking up files that aren't supposed to become assets inside Final Cut Server. If you're going to make your scans manually through the Administration window of the client application, it's recommended that you mimic this Exclude Filter; otherwise, your Final Cut Server might fill up with useless assets.

NOTE ▶ The reason you exclude ".*" is to avoid accidentally adding any Apple double files to the catalog. You also exclude "Thumbs.db," as that file is created by Windows users browsing subdirectories and is useless if added to the catalog.

You can't modify these values via System Preferences, so if you want to include only a certain file extension (as you'll do in the next exercise), you need to create your scan through the Administration window of the client application.

4 Click Discard Changes and click Yes to confirm. Close the Administration window, click Don't Save, and then log out of the client application.

ADMIN ▼

Discovering Video Assets Using Scans

Now that you've seen the differences between adding assets via System Preferences and via the Administration window, let's clean up the old device and recatalog it via a set of scans tailored to specific types of media and metadata.

Before you proceed with creating the new filtered scans, you need to clear out what you did in the previous exercise. The easiest way to do this is to delete the device from the Final Cut Server pane in System Preferences. This will remove any assets and responses associated with this device from the catalog but will not touch the high-resolution material that resides in the Evergreen folder.

1 Open System Preferences, choose Final Cut Server, and authenticate.

2 At the top of the Final Cut Server pane, click Devices, select Evergreen, and click the Delete (–) button to remove the device and any assets, metadata, and automations associated with it. Read the warning carefully, and then click OK.

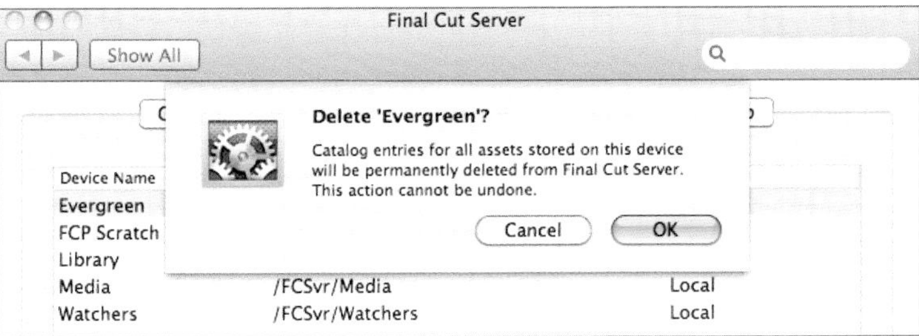

3 Repeat steps 4–10 from the "Setting Up a Basic Scan" exercise, but this time leave the Full Scan checkbox deselected when proceeding through the Device Setup Assistant. Here's a short list of steps:

- Create a new device.
- Device Type: Local
- Device Name: *Evergreen*
- Location: Macintosh HD/FCSvr/Evergreen
- Archive Device: No
- Scan Settings: Default Off
- Transcode Settings: No Conversion

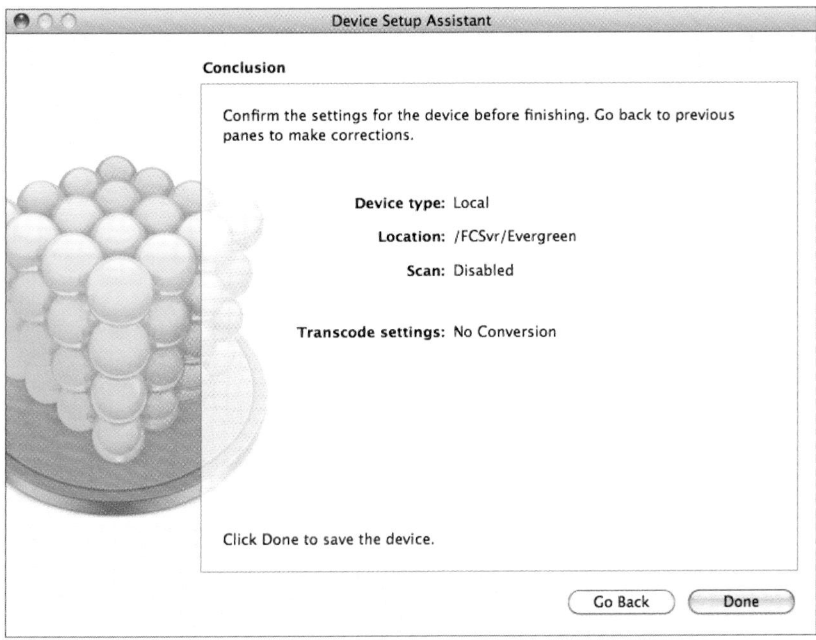

4 Quit System Preferences.

5 Log in to the Final Cut Server client application as administrator, and open the Administration window.

6 At left, click Response, and then click the Create (+) button to add a new response.

7 In the Response window, choose Scan from the Response Action pop-up menu at the top; and in the Name and Description fields, enter *Evergreen – Full Scan [Video]*.

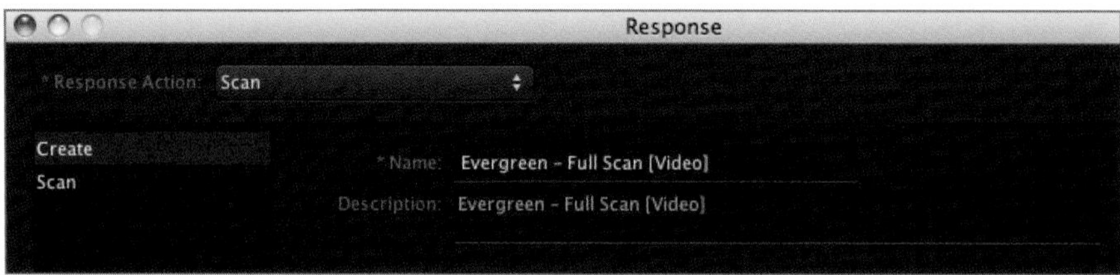

8 On the left, choose Scan, and choose or enter the following values:

- Scan Source (the device to be cataloged): *Evergreen*
- Metadata Set: Media
- Scan Mode (type of scan): Full
- Entity Type (type of asset to catalog): File
- Video Frame Rate (if it is a video asset): Auto
- Recursion Limit (how many subdirectories to include, 0=infinite): 0
- Wildcard Include Filter (the file extensions to pick up): *.mov
- Wildcard Exclude Filter (the file extensions to ignore): .* and Thumbs.db.

TIP For the filters, enter the value in the field, and then click Add. To remove a filter, select the filter in the list, and then click Remove.

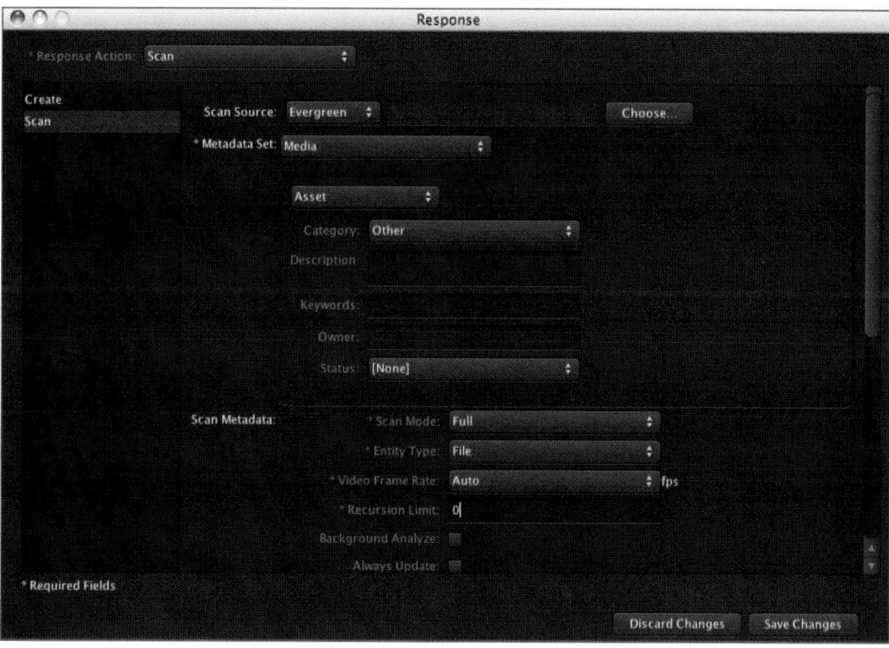
Response
* Response Action: Scan
Create
Scan
Scan Source: Evergreen
Choose...
* Metadata Set: Media
Asset
Category: Other
Description:
Keywords:
Owner:
Status: [None]
Scan Metadata:
* Scan Mode: Full
* Entity Type: File
* Video Frame Rate: Auto fps
* Recursion Limit: 0
Background Analyze:
Always Update:
* Required Fields
Discard Changes
Save Changes

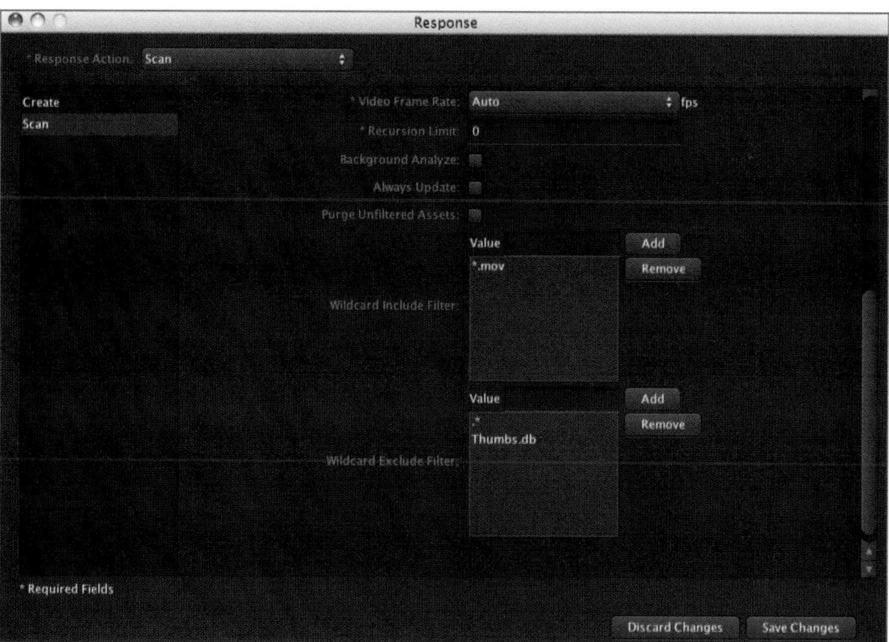
Response
* Response Action: Scan
Create
Scan
* Video Frame Rate: Auto fps
* Recursion Limit: 0
Background Analyze:
Always Update:
Purge Unfiltered Assets:
Value
Add
*.mov
Remove
Wildcard Include Filter:
Value
Add
.*
Thumbs.db
Remove
Wildcard Exclude Filter:
* Required Fields
Discard Changes
Save Changes

9 After reviewing the screen shots and comparing your selections to make sure they are accurate, click Save Changes.

Now you have to manually set up a schedule for the scan you just created.

10 In the Administration window, click Schedule, and then click the Create button (+) to make a new schedule.

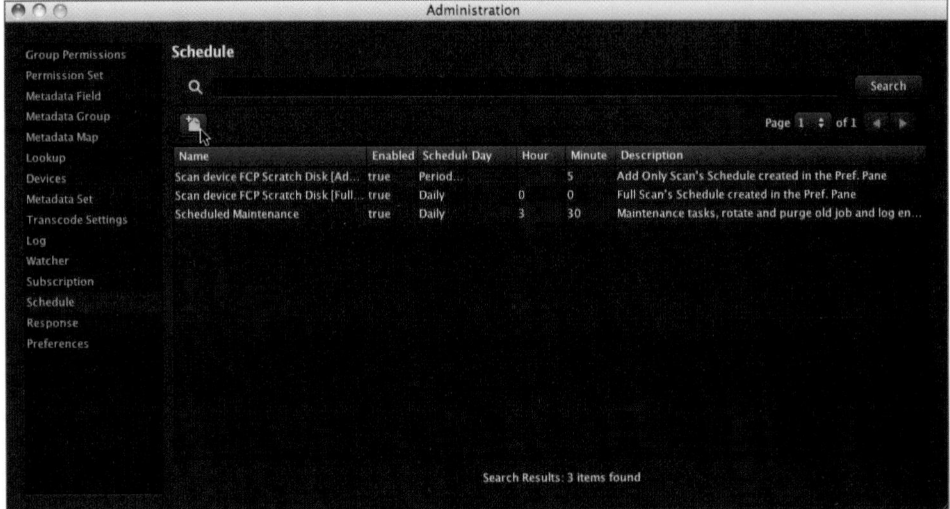

11 At the top of the Schedule window that opens, choose Periodically, and make sure that the Enabled checkbox is selected. Enter *Evergreen - Full Scan [Video]* in both the Name and Description fields.

12 Choose "Evergreen — Full Scan [Video]" (the response you made earlier) from the Available list on the right, and click Add to add it to the Selected list on the left.

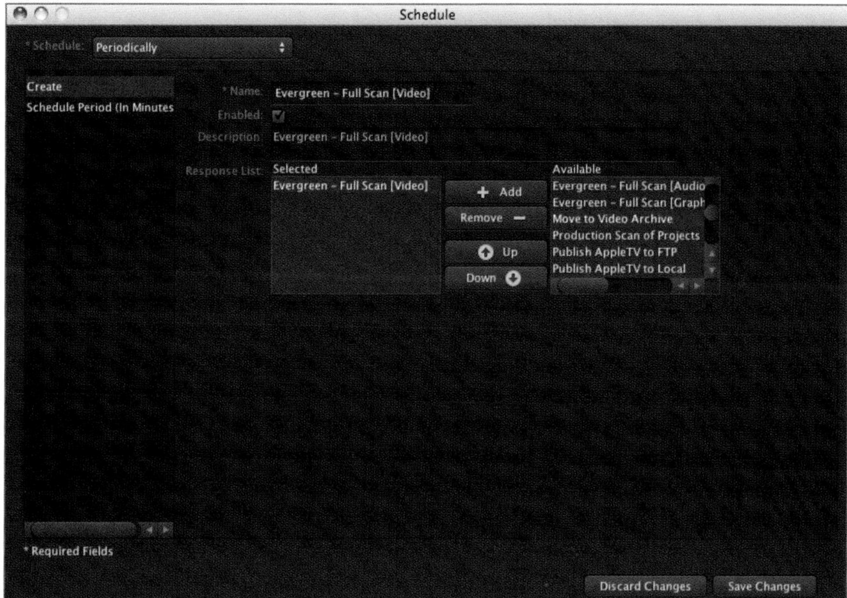

13 On the left, choose Schedule Period (in Minutes) and enter *3* for this exercise.

This will fire off a full scan every three minutes looking for new assets, deletions, and any file modifications.

NOTE ▶ In a real-world scenario with hundreds or thousands of different assets and file system and network scenarios, you would not want to run a full scan every 3 minutes. The add-only scan picks up new material and should be utilized in place of full scans.

14 Click Save Changes, and then close the Administration window.

After three minutes, your full scan should kick off and start cataloging any content that is on the Evergreen device that ends in *.mov* (the video files). You can monitor the proxy creation in the Search All Jobs window and verify that all of the content has been created correctly by doing a search in the client application with no keywords. This will show you the entire catalog of assets in Final Cut Server.

15 Click Refresh to update the Assets pane.

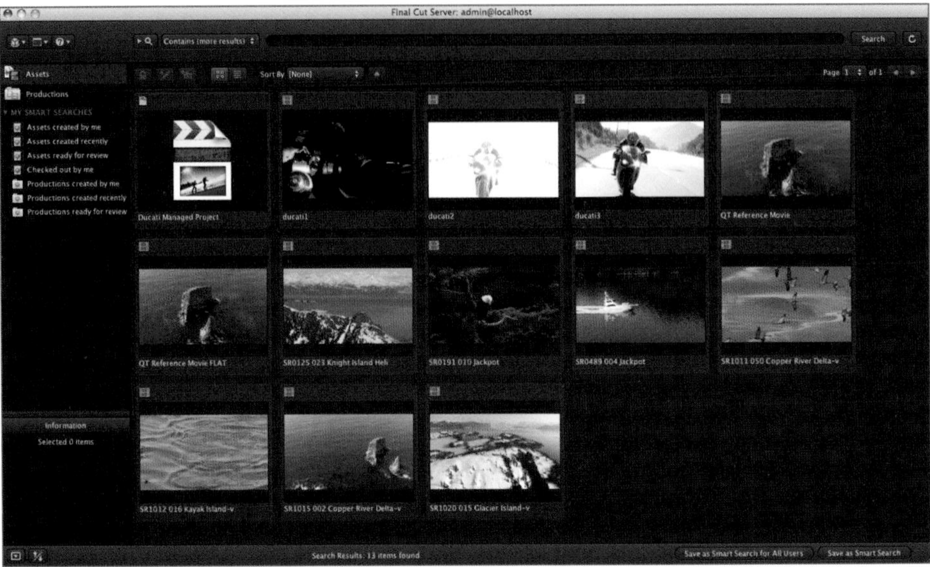

Notice that the scan response has picked up only video assets and that the other files in the directory have been left uncataloged.

16 Double-click the asset icon of **SR0125 023 Knight Island Heli.**

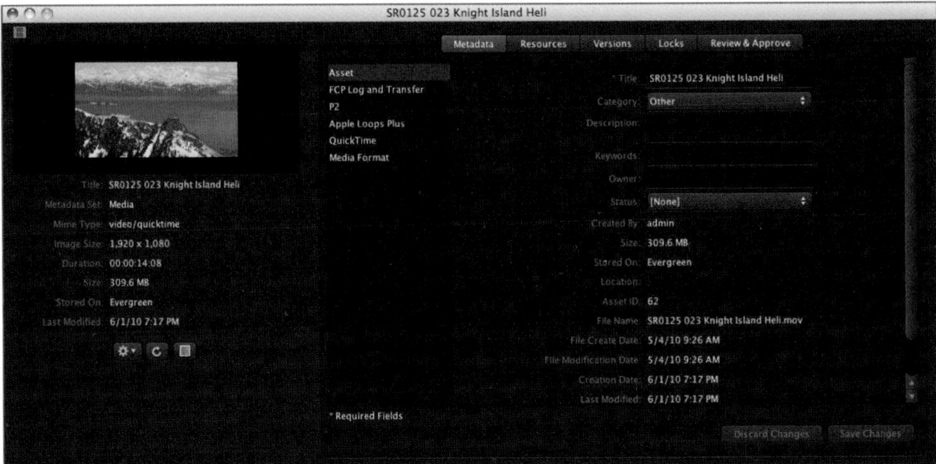

Notice that the Media metadata set with its associated groups and fields were applied to this asset.

17 Close the asset's info window.

ADMIN ▼

Discovering Audio Assets Using Scans

You're going to take what you learned in the previous exercises about using scans to discover video assets, and apply the same methodology to discovering audio assets.

The main advantage you gain from having two separate scan processes for audio and video (and later in this lesson, graphic assets) is that you can have different metadata automatically stamped on the assets, and you can apply different metadata sets to the assets.

You're going to create a new scan response and a new schedule for that scan. In very small environments where you don't have many assets, you can use the same schedule for your scans. If you have a larger facility, it is recommended that you stagger the scheduling of your scans so that they overlap as infrequently as possible. If scans overlap, asset creation gets delayed until the proxy-creation events from the previous scan have finished, thereby making the periodic discovery of assets unreliable.

1 Open the Administration window.

2 Choose Response from the choices on the left.

   You're now going to duplicate the response you made in the previous exercise.

3 Highlight the Evergreen – Full Scan [Video] response, and click the Duplicate button.

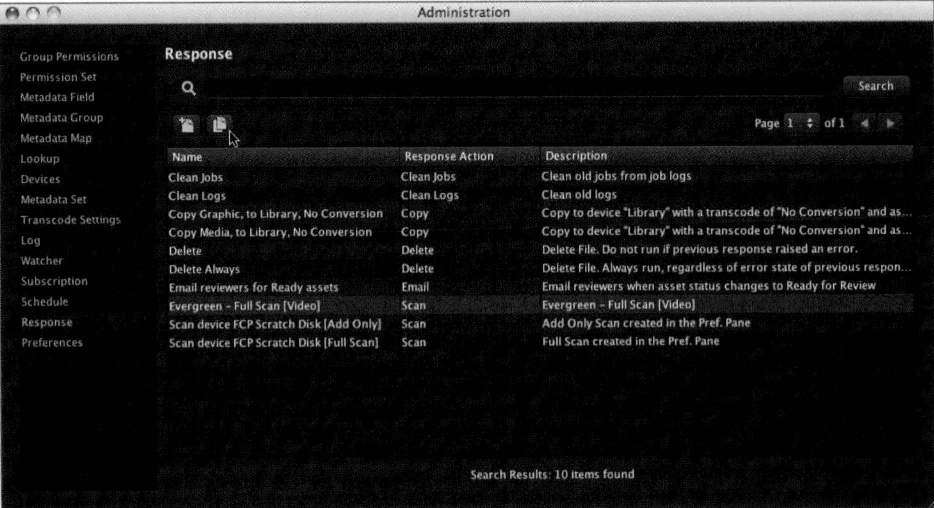

4 Double-click the newly created Clone of Evergreen – Full Scan [Video].

5 In the Name and Description fields, enter *Evergreen – Full Scan [Audio]*.

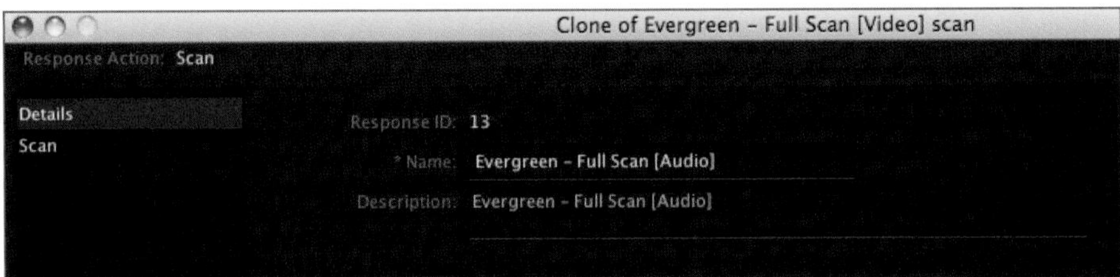

6 Under Scan, leave Scan Source the same, but change Metadata Set to Audio to reflect the new audio assets you are about to catalog.

7 Remove *.mov from the Include filter. Add **.aif* to reflect the extensions of the audio assets. Click Save Changes.

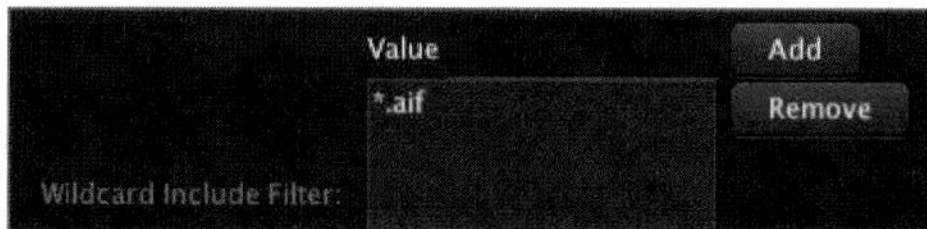

8 In the Administration window, click Schedule. Click the Create button to create a new schedule.

9 Choose Periodically from the Schedule pop-up menu, and give the schedule a relevant name and description such as *Evergreen — Full Scan [Audio]*.

10 Select Enabled.

11 Choose the Evergreen — Full Scan [Audio] response you made from the Available list on the right, and add it to the Selected list on the left.

12 Set Schedule Period to 2 minutes. Click Save Changes.

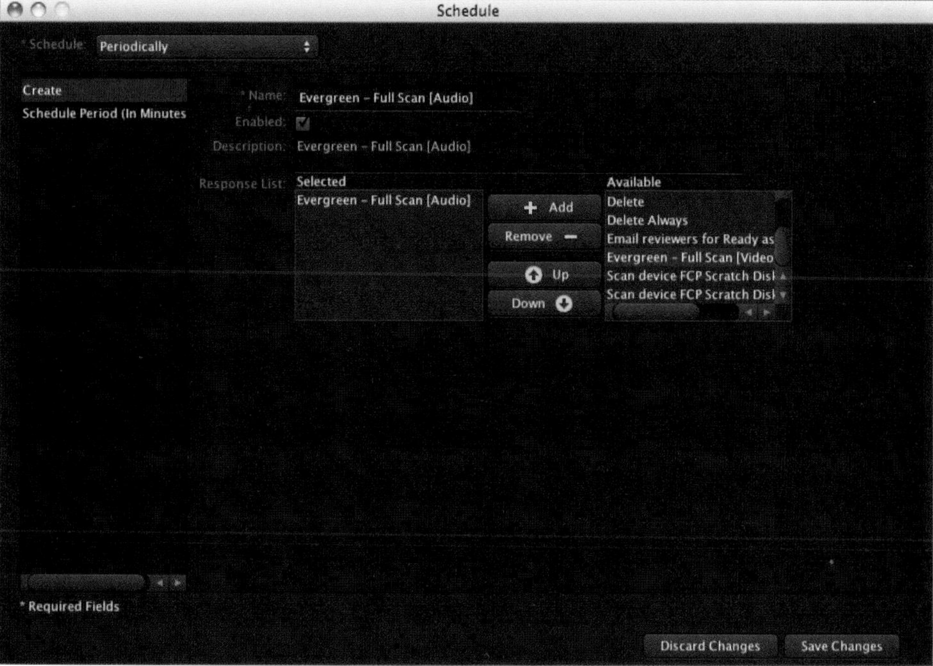

After two minutes, your scan will run and pick up your audio assets from the Evergreen device.

13 Close the Administration window, and then refresh the Asset pane.

14 Double-click the **42 Rfx** asset.

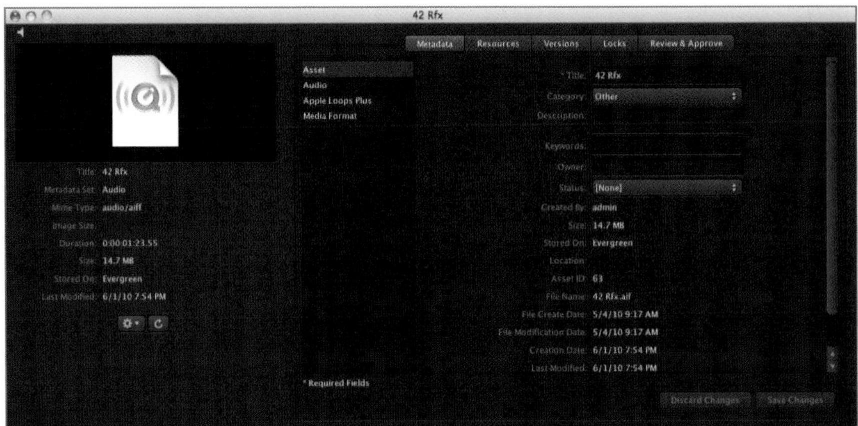

Notice that it's using different metadata from the video files you cataloged earlier. This is because you chose the Audio metadata set for the second scan in order to apply different metadata to the audio files as opposed to the video files.

15 Click the Apple Loops Plus metadata group.

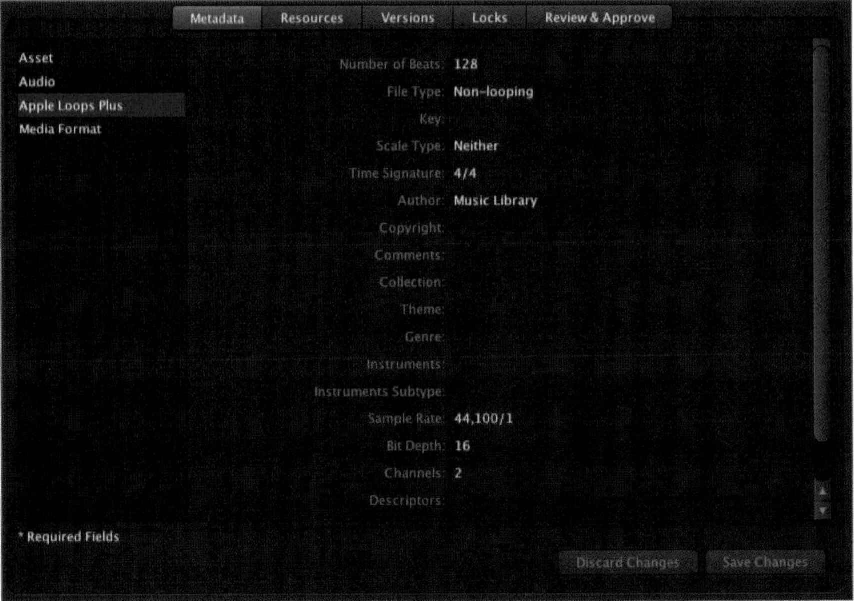

Notice the audio-specific metadata that has been automatically mapped into Final Cut Server from the information in the audio file. This functionality is available only with the proper metadata set (Audio).

NOTE ▶ To get the Apple Loops Plus metadata information, it's not necessary to use the Audio metadata set (although it's the simplest and only built-in way to use that metadata). You can create your own custom metadata sets and add the Apple Loops Plus metadata group to them, and then your audio metadata will be automatically accessible.

ADMIN ▼

Discovering Graphic Assets Using Scans

Now that you've learned how to add your video and audio assets to Final Cut Server, you'll apply that knowledge to discovering graphic assets. Just like audio and video assets, graphic assets also have their own metadata set: Graphic. By utilizing separate scans, you can specify that your graphic assets get the Graphic metadata set applied, so that the relevant metadata fields will be present.

Final Cut Server can work with most common graphical file types, including RAW, TIFF, Targa, bitmap (BMP), JPEG, and Photoshop (PSD). It also supports reading and writing IPTC and XMP metadata to and from the graphic assets. For now you'll concentrate on simply getting the assets into the catalog.

You're going to create a new scan response specifically targeting graphic files and a new schedule for that scan. In very small environments where you don't have many assets, you can use the same schedule for your scans. If you have more than one edit station, it is recommended that you stagger the scheduling of your scans so that they overlap as infrequently as possible.

The first thing you are going to create is a new scan response.

1 Open the Administration window.

2 Choose Response from the choices on the left.

3 Select the Evergreen – Full Scan [Video] response, and click the Duplicate button.

4 Double-click the newly created Clone of Evergreen – Full Scan [Video].

5 Change the Name and Description fields to *Evergreen – Full Scan [Graphic]*.

6 Under Scan, change Metadata Set to Graphic to reflect the new graphic assets you're about to catalog.

7 Remove *.mov from the Include filter. Insert **.tga* to reflect the extensions of the Targa graphic assets. Click Save Changes.

NOTE ▶ The reason you added only .tga to the Include filters is because the graphic example for this exercise is a Targa file. If you were working with multiple file formats (JPEG, TIFF, BMP, and so forth), you would need to include those wildcard filters as well (*.jpg, *.tiff, *.bmp, and so on).

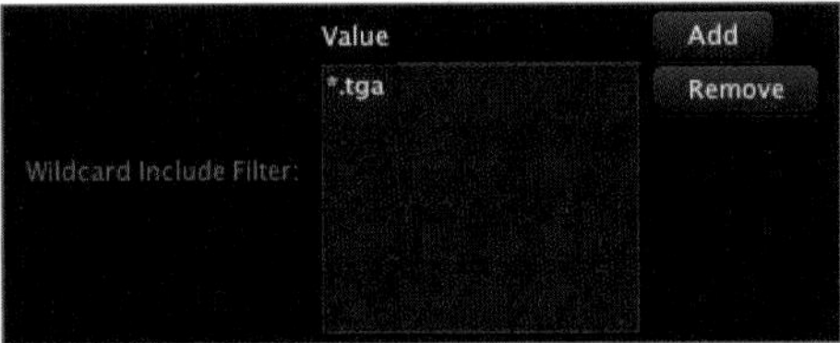

8 In the Administration window, click Schedule. Click the Create button to create a new schedule.

9 Choose Periodically from the Schedule pop-up menu, and change the Name and Description fields to *Evergreen – Full Scan [Graphic]*.

10 From the Available list on the right, choose the Evergreen – Full Scan [Graphic] response you made, and add it to the Selected list on the left.

11 Select Enabled.

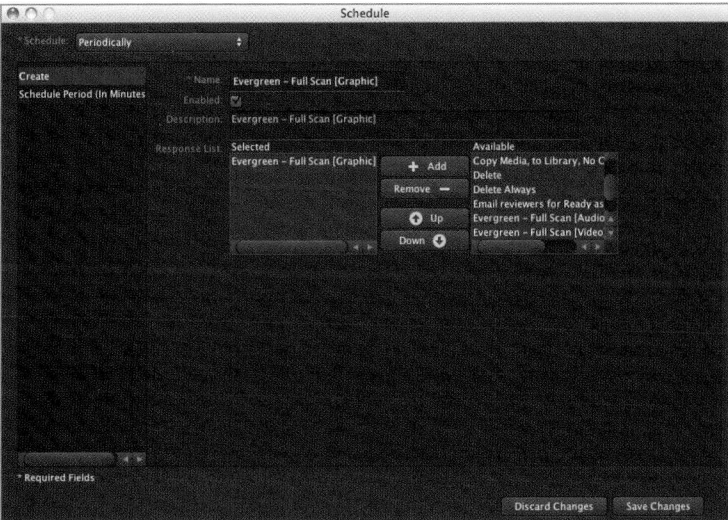

12 Set Schedule Period to 2 minutes, and click Save Changes. Close the Administration window.

After two minutes, your scan will run and pick up the graphic asset from the Evergreen device.

13 Refresh the Asset pane.

14 Double-click the asset **Evergreen Logo – (Keyable)** to bring up the metadata record for that asset.

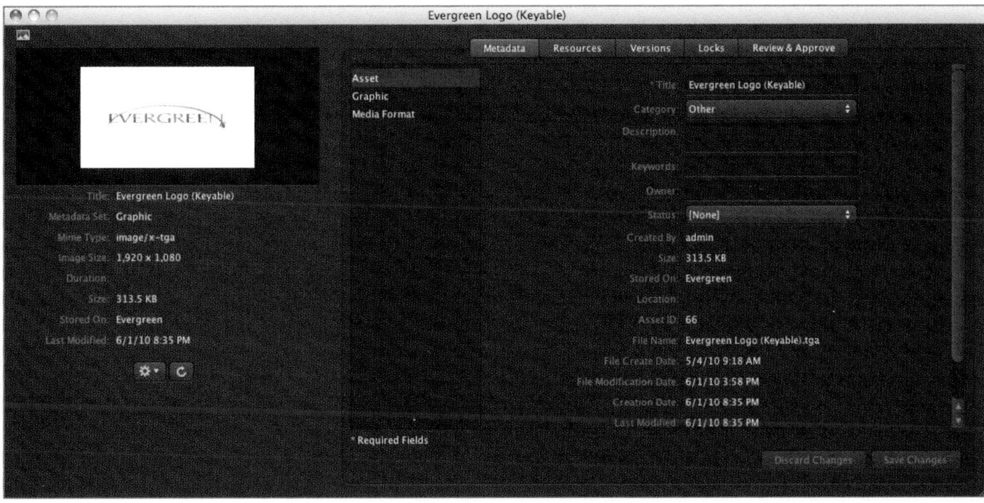

Notice that it's using different metadata from the video and audio files you cataloged earlier. This is because you chose the Graphic metadata set for the third scan in order to apply different metadata to the graphic files, as opposed to the video and audio files.

15 Close the asset's info window.

ADMIN ▼

Discovering Assets Using Watchers

In the previous exercises, you learned how to utilize the Final Cut Server scan functionality to add different types of assets into the catalog. Another way to get assets into Final Cut Server is by utilizing watchers.

Watchers are folders that are on a device that is constantly being watched for new files. Ninety percent of the time, this is done so that the file can be moved and possibly transcoded to another device.

A common scenario involves copying the file directly to another device, such as a network-mounted, shared, or Xsan file system, and from there to the Final Cut Server catalog.

You'll make three different watchers in this exercise: Audio, Video, and Graphic. Each will add its specific file type to the Final Cut Server catalog.

First you're going to make the folders in the Finder that Final Cut Server is going to be watching. During the installation, Final Cut Server makes a default device called Watchers. It's located at the root of where you installed the Final Cut Server proxies and media.

1 Open the Finder and navigate to Macintosh HD/FCSvr.

NOTE ▶ In the real world, you would find this on a separate server that is sharing the watchers via AFP (or some other network protocol, or a mounted file system such as Xsan). For this exercise, this folder is located on your local system in FCSvr.

Inside the Watchers folder you'll notice two default folders: Graphic and Media. Final Cut Server comes with two built-in automations; you'll utilize the Graphic watcher shortly. But first, you need to create one new folder.

2 Create a new folder named *Audio* inside the Watchers folder.

These folders will be only a temporary storage point. You'll have Final Cut Server take the files copied into the folders and then push them to a separate device.

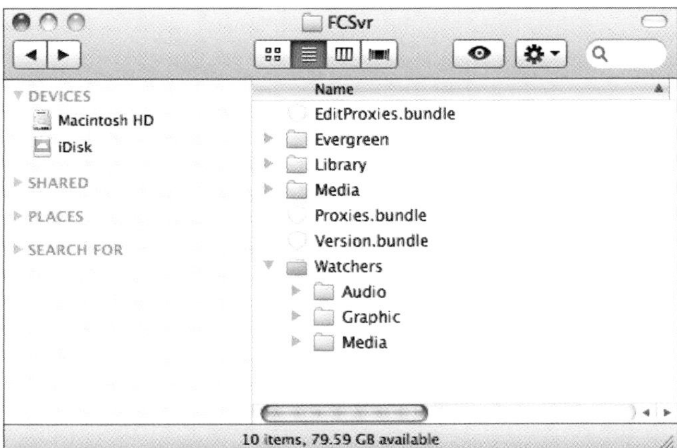

3 Open System Preferences and click the Final Cut Server icon. Authenticate yourself as the administrator.

4 Click the Automations tab.

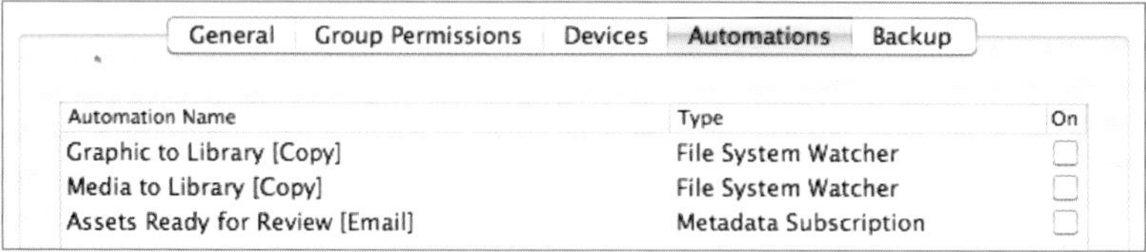

Notice the existing automations that are installed by default. You'll be utilizing both the default Media and Graphic automations as they mimic the behavior you created earlier with the scans (except that in this case, you're copying a file from your desktop to another device, whereas a scan leaves the file in place).

5 Click the Add (+) button to add a new automation.

6 In the Automation Setup Assistant, select File System Watcher and click Continue.

NOTE ▶ You'll touch on metadata subscriptions in later lessons.

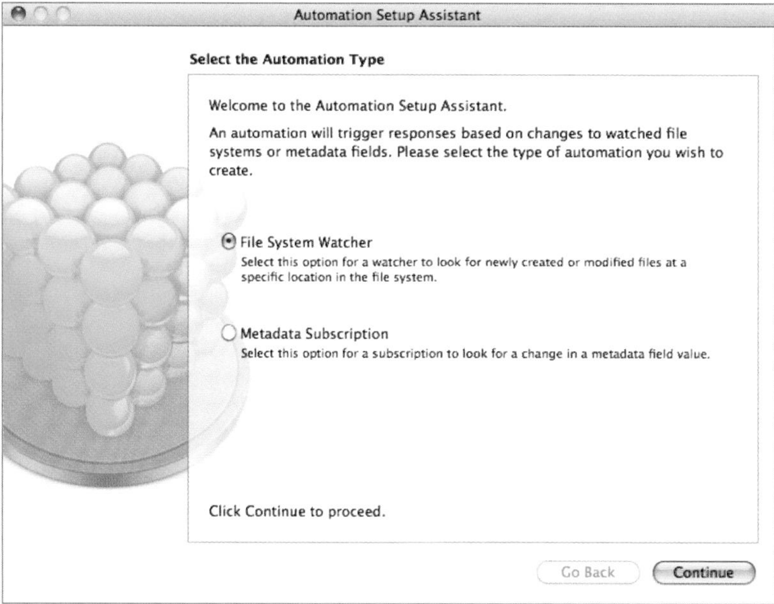

7 Name the automation *Audio to Library*, choose Watchers from the Device pop-up menu, and select the Watch Subfolder checkbox. Click Browse and select the new Audio subfolder you created. Click OK.

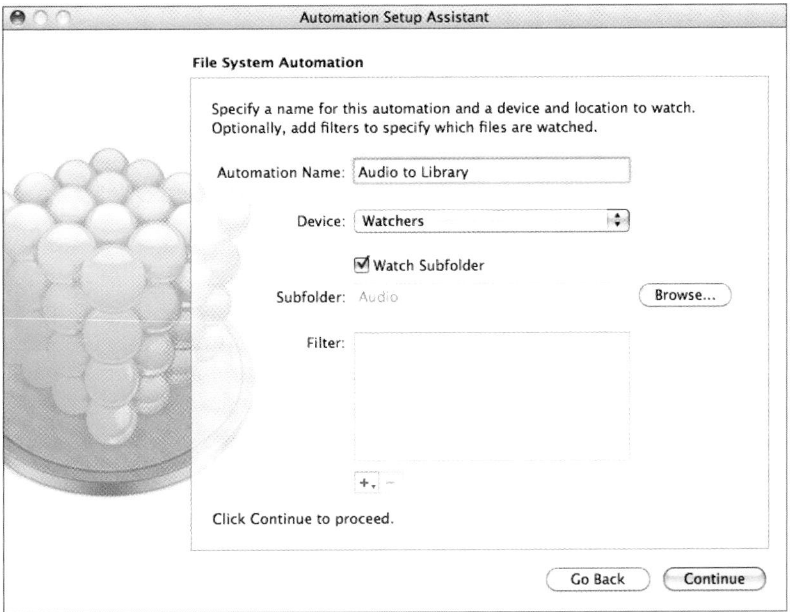

8 Click the Add (+) button to add a new filter, choose Custom, and enter **.aif*. Click OK.

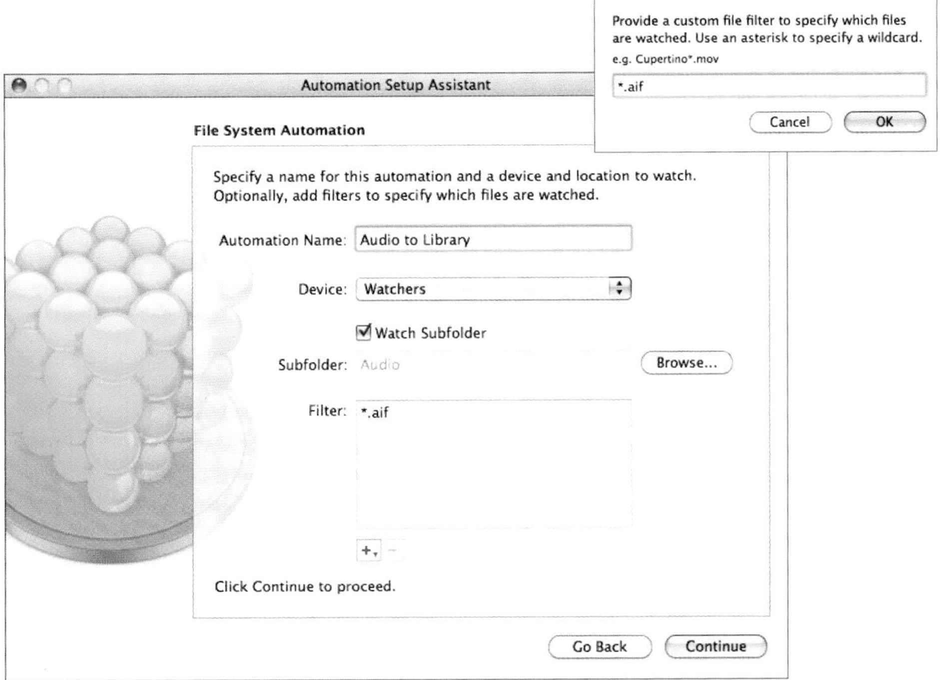

9 Click Continue.

In the next dialog, you'll define the responses that Final Cut Server will do to the discovered files.

10 Click the Add (+) button in the bottom left of the dialog, and choose Copy Response. Choose Library from the Destination Device pop-up menu, and choose Audio from the Metadata Set pop-up menu. Later we'll talk about transcoding with watchers, but for now you'll leave Transcode set to No Conversion.

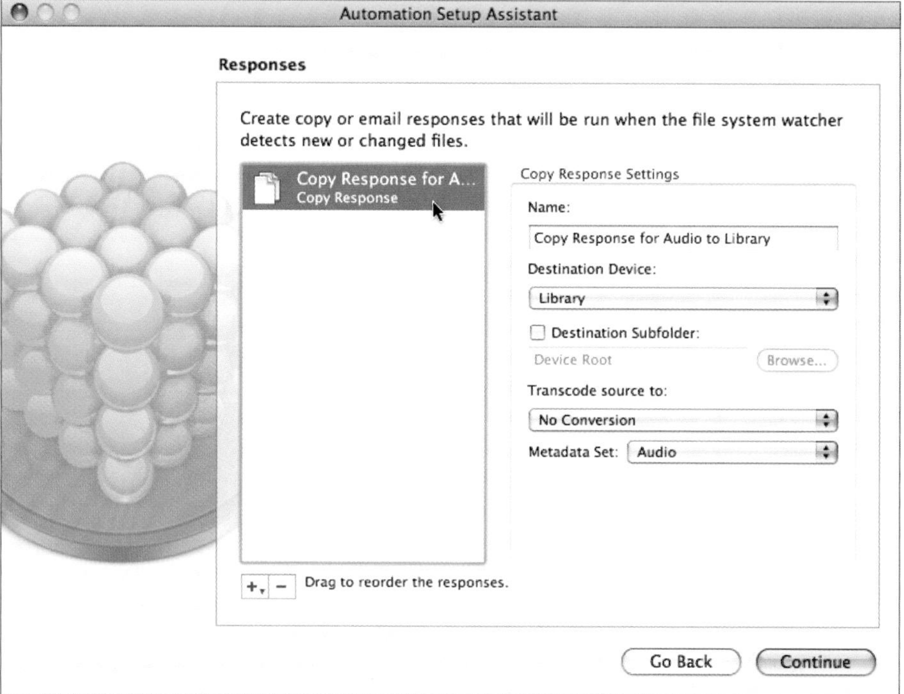

11 Click the Add (+) button in the bottom-left corner and choose Delete Response.

TIP You can reorder the responses by dragging them, but make sure that your delete response is the last response in the list.

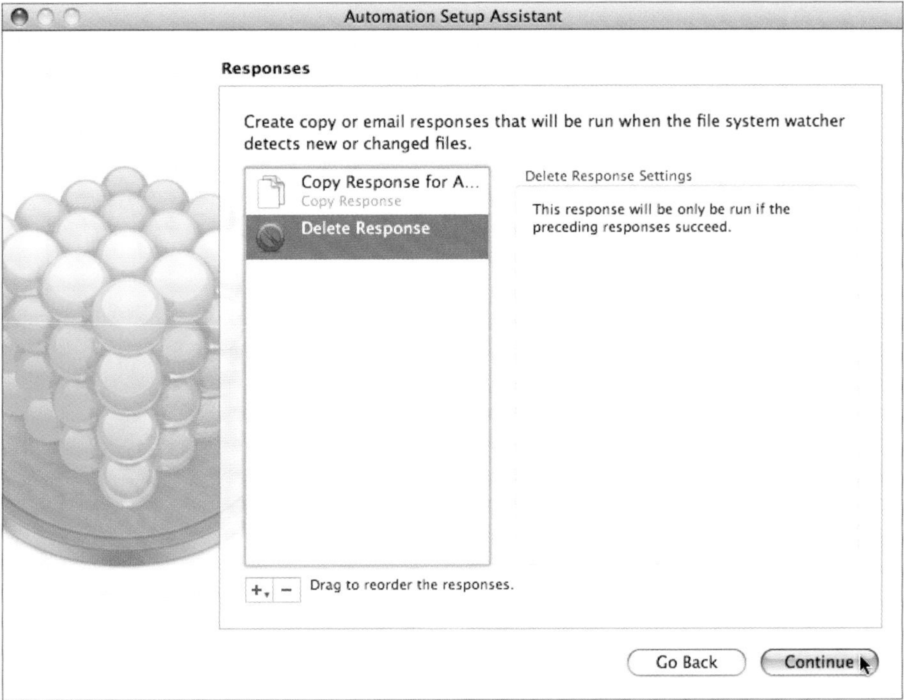

The delete response will remove the file from the watcher on the local machine, after it has been copied to the Final Cut Server catalog and the Library device.

12 Click Continue.

13 In the next window, you have a chance to review the watcher automation you've created and make sure that everything has been configured correctly. When ready, click Done. Final Cut Server returns you to the main Automations pane in System Preferences.

14 Back in the main Automations pane, enable the Graphic to Library [Copy] and Media to Library [Copy] automations by selecting the checkboxes under the On column to the far right of the names.

Turning on these automations tells Final Cut Server to begin watching the folders for new files. When new files hit the folders, the responses you associated with the watcher will be triggered.

15 Quit System Preferences.

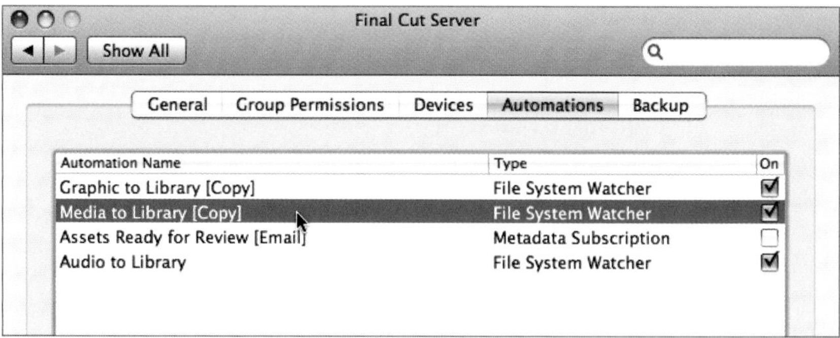

USER ▼

Using Sample Media to Trigger the Watchers

Now that you have configured Final Cut Server to utilize both the new and default watchers, you'll use sample media from the included FCS_Book_Files to trigger these watchers.

You'll be copying the media to the specific watchers to match their type and extension. This will trigger the copy actions and push the material (without a transcode) to your Library device. The material will then automatically be removed from the watchers by the delete response you added after the copy response.

1 Open the Finder and navigate to the FCS_Book_Files folder.

2 Open the folder For Watchers – No Transcode. You'll see four files: one graphic, one video, and two audio.

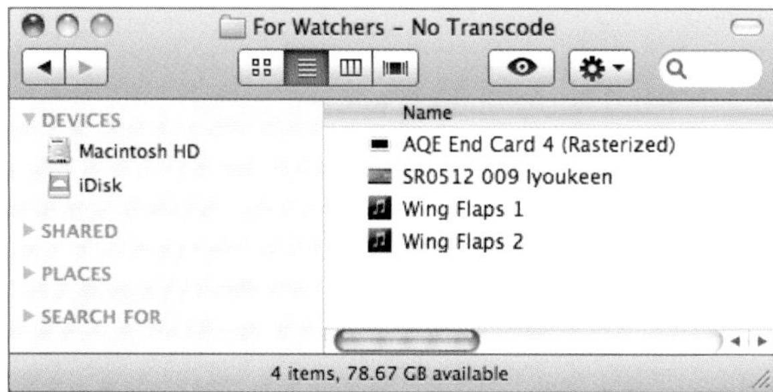

3 In another Finder window, open the folder containing the Macintosh HD/FCSvr/Watchers folder.

4 Drag the **Wing Flaps 1** file into the Audio folder.

5 In the client application, open the Search All Jobs window to monitor the progress of the watcher.

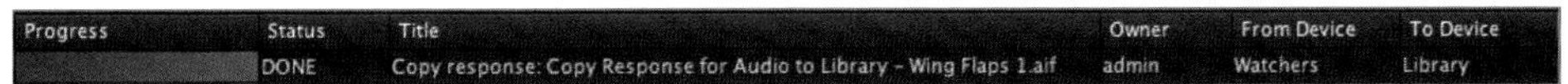

After a few moments, you'll notice that the file is copied to the Library device.

6 Drag the **AQE End Card 4 (Rasterized)** file to the Graphic folder.

Again, notice in the Search All Jobs window that the file has been copied to the Library device.

7 Drag the **SR0512 009 lyoukeen** file to the Media folder and see that the file has been copied to the Library.

8 Refresh the Asset pane to verify that all three assets have been added to the catalog.

Fortunately, watching these folders and adding their contents to the catalog was not the only response action you assigned.

9 Navigate to your watchers at Macintosh HD/FCSvr/Watchers. Open each of the three enclosed folders.

   These folders are now empty because each response included the delete action. Ending a response with delete ensures that files are not duplicated within the catalog when the watcher activates.

In this lesson you learned how to use the Final Cut Server built-in automation tools to add files to the catalog as assets. Scans are powerful when you have an existing directory structure, as that structure becomes searchable metadata in Final Cut Server and thus you don't have to move your media from its original location. Scans also offer you the ability to segment your content, based on metadata set, file extension, or physical location.

You'll use watchers again later when it comes time to distribute the finished assets. But first you'll explore ways to search and organize your assets.

Lesson Review

1. What are the differences between add-only and full scans?
2. How do you specify a file filter for a scan?
3. True or false: Scans are the only way to ingest material.
4. How often should you schedule full scans?
5. Should you use a scan or a watcher when attempting to transcode material?

Answers

1. Add-only scans pick up only files with a create date newer than the last time the scan was run. Full scans pick up new assets, modifications to existing assets, and deletions of assets from the file system being scanned.
2. To specify a filter for a scan, you need to create the scan in the Administration window of the client application. This option is not available to you in System Preferences.
3. False. You can also use watchers to ingest material, or you can simply drag it to the Final Cut Server client application.

4. On a production, full scans should only be scheduled daily. If you don't have a large amount of material (or bandwidth concerns), you can schedule full scans twice daily.
5. If you're going to be transcoding material, the easiest way to achieve this is with a watcher.

5

Goals

Search for assets

Modify the advanced search filter

Save searches

Delete assets

Lock assets

Create productions

Lesson **5**

Searching and Organizing Assets

As you create thousands, tens of thousands, and hundreds of thousands of assets, the metadata you enter during ingest gives you the ability to quickly find the desired assets. The metadata combinations possible for each asset are mind-boggling. Yet, Final Cut Server is built to make searching that complex metadata simple and fast.

Final Cut Server allows you to modify the interface in ways to speed up searches. These modifications allow you to customize advanced searches by any of the metadata fields. And after you have found the desired assets, you can save the search parameters for quick recall.

When you have found assets you need to track together, you can store those assets in a production—a virtual folder of assets that can be easily shared with other users. Any asset type may be added to a production, making the productions into extensions of multiple searches.

You'll get started by looking at available search parameters and then move into the advanced search parameters.

USER ▼

Structuring Search Queries

Searching for assets is one of the most powerful features of Final Cut Server. The more specific your search criteria, the more meaningful the search result will be. Don't forget though, that search is meaningless if metadata has not been applied to the assets.

To the left of the search field, you'll find the Search pop-up menu. This menu affects the speed of the search. The Search pop-up menu has two options: Contains and Matches Word. In either case, Final Cut Server searches through all metadata for your keywords. Don't worry about case sensitivity, because it's ignored in searches.

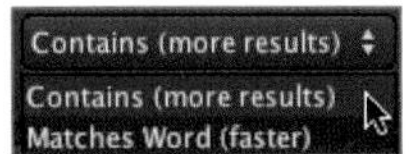

Choosing "Contains (more results)" returns results in which the keyword matches or is contained in a larger word. This option means a slower search but yields more results.

Choosing "Matches Word (faster)" returns only results in which the entire keyword matches. This means a faster search but yields fewer results.

You can search for multiple keywords by typing them together, separated by spaces. Assets whose metadata contains all the typed keywords will be returned.

USER ▼

Viewing Search Results

Search results appear in the lower portion of the Viewer. You can change how you view the results with the view buttons.

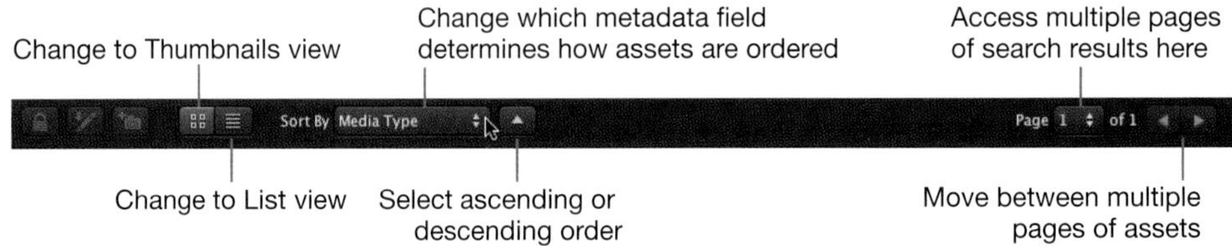

Your search might return more results than can be seen in one page. You can use the page-navigation controls in the upper-right corner of the window to view additional pages of results.

You can also view additional details about a selected asset with the Show Item Information button, in the lower left of the window.

In all views, you will see an identifying icon that visually denotes the kind of asset you are viewing. Clicking this icon when viewing a search result in Thumbnails view allows you to preview visual assets, such as images and video clips.

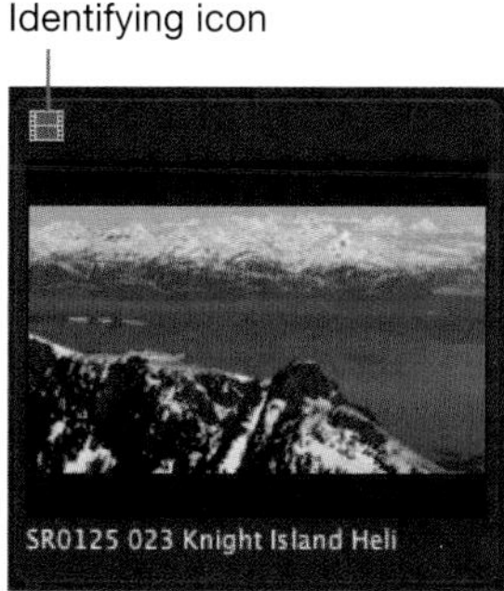

NOTE ▶ You can preview assets, such as images and some audio and video clips, while viewing a search result in Thumbnails view. Other kinds of assets, such as Final Cut Pro project files, PDFs, and other text-based documents cannot be previewed. You must add them to your cache to use them.

Final Cut Server assigns the following icons to assets depending on their type:

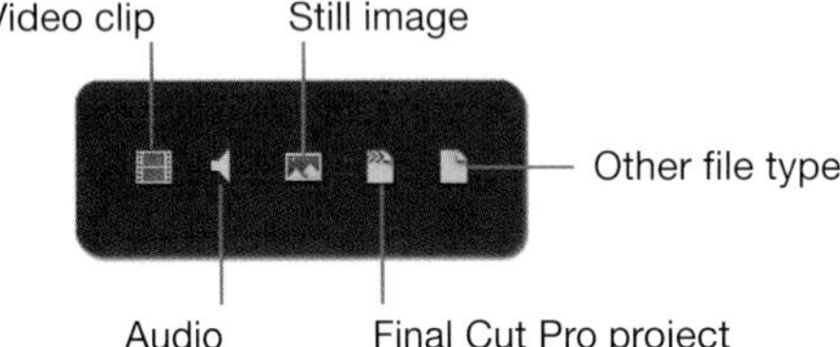

In some cases, you may see one or more additional icons to the right of the identifying icon that visually describe the state of the asset.

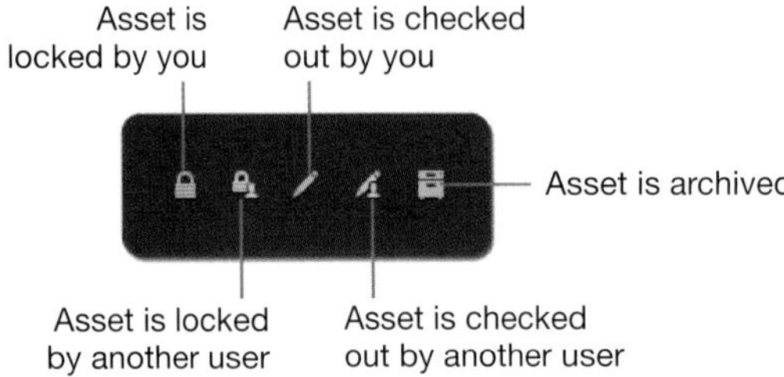

USER ▼

Adjusting Search-Results Settings

The Final Cut Server client application has limits on the maximum number of results that are returned and the maximum number of results per page.

To change these and other settings, choose Preferences from the Server pop-up menu in the upper-left area of the client application window.

If you select the "Remember last search" checkbox, the next time you log in, Final Cut Server will automatically display the last search you performed.

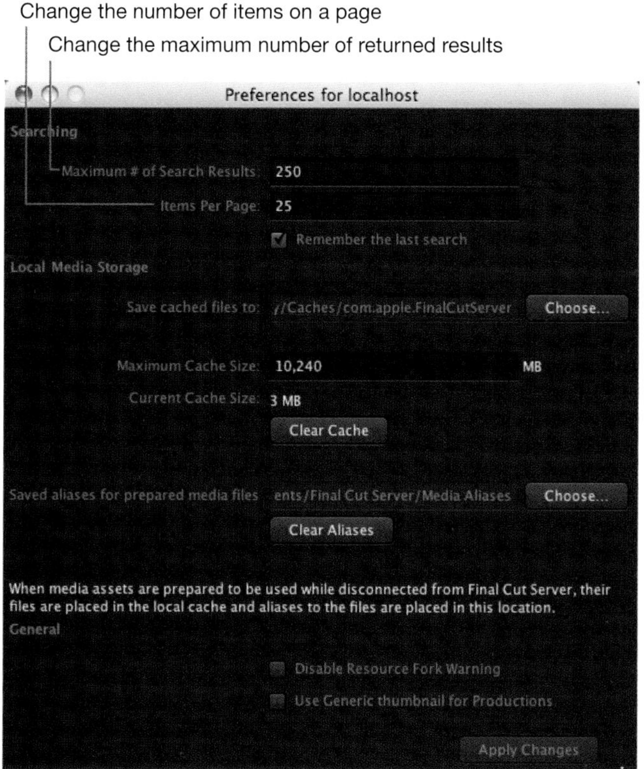

ADMIN ▼

Customizing Advanced Searches

As an administrator, you can customize the metadata structure applied to every asset within Final Cut Server. As users upload new assets, they're tasked with entering the required metadata so that future users may find the assets quickly. As you have seen so far and will cover in more detail later, the metadata applied may go well beyond simple keywords.

The Advanced Search panel helps users sift through the mountains of metadata with ease; however, the administrator must customize the panel to allow users to search the customized metadata. The Asset Filter metadata group is a special group used by Final Cut Server to filter individual fields in an advanced search. An administrator may modify this metadata group to alter the Advanced Search panel.

While uploading assets in an earlier lesson, you selected the Good checkbox in the Log and Transfer window. By default, this metadata morsel is not a specific parameter that can be used when searching the catalog. You will set Final Cut Server to allow searching for all clips that are tagged as Good.

1 If necessary, open the client application and log in as a Final Cut Server administrator.

2 Open the Administration window. Choose Metadata Group from the list on the left, and enter *Asset Filter* in the search field. Double-click Asset Filter in the search results.

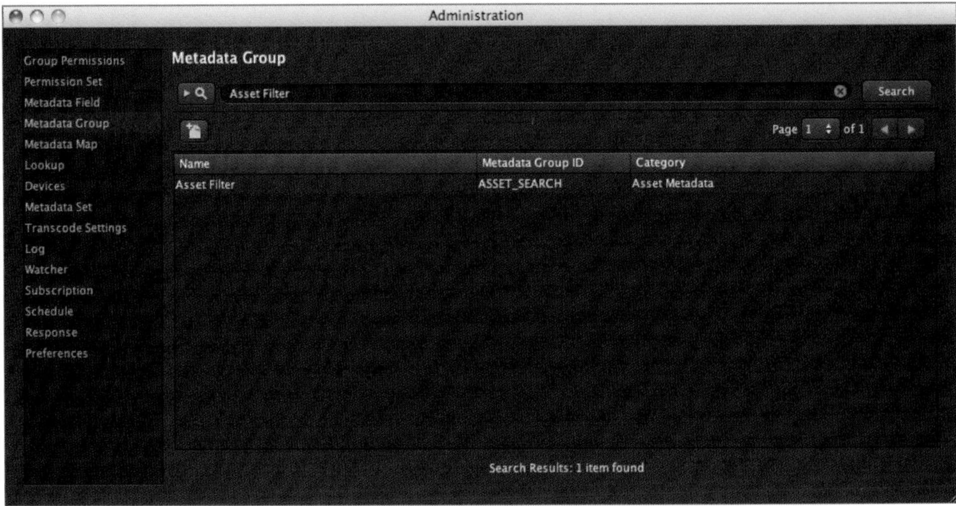

The information window for the Asset Filter metadata group opens. The top boxes are for adding individual metadata fields to the group, with Available fields on the right, and those that are already associated with the metadata group on the left.

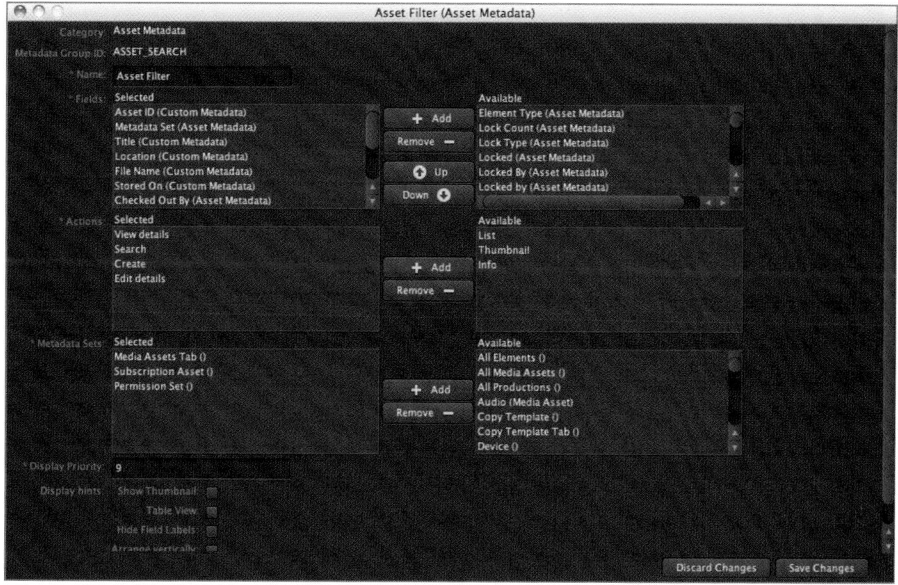

3 Select IsGood (Custom Metadata) from the Available list on the right, and click Add to add it to the Selected list on the left.

4 Verify that IsGood has been added correctly by scrolling down in the Selected field list, and then click Save Changes.

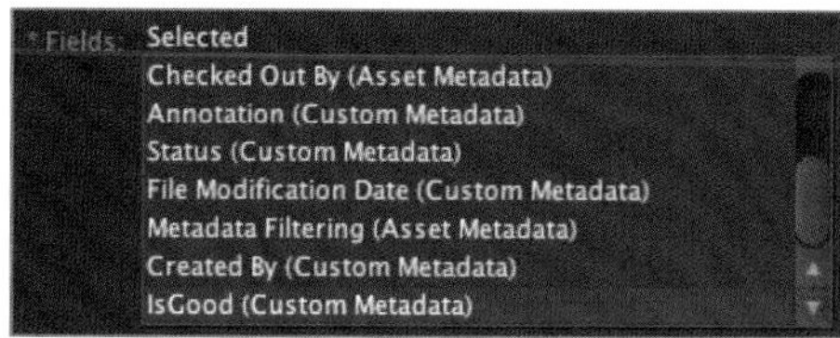

5 Close the Administration window.

Before any user can utilize a metadata group modification an administrator has made, the client application must be refreshed.

6 Click the Window pop-up menu and choose New Workspace, and then close the old workspace window.

USER ▼

Utilizing Advanced Search and Creating Smart Searches

Now that you've configured the Asset Filter metadata group, you'll search for assets that were marked as Good during ingest within Final Cut Pro. Then, you'll be using the Smart Search function so that you may recall the same search again at a later time.

1 Click the disclosure triangle next to the magnifying glass at the top of the window to expand the advanced search options.

Notice that IsGood is now an option to narrow your search results.

2 Choose True from the IsGood pop-up menu in the advanced search filter, and then click Search.

Notice that your **ducati1** asset comes up because you selected the Good checkbox in the Final Cut Pro Log and Transfer window. That information has been automatically mapped to Final Cut Server.

In the bottom right of the client application window, there are two buttons: "Save as Smart Search for All Users" and "Save as Smart Search." Both buttons are visible only when you're logged in as an administrator. Non-admin users will see only the "Save as Smart Search" button. As the button name implies, the administrator version allows you to create a smart search that is global to all users.

3 Click "Save as Smart Search" and name the search *Good Clips.*

This smart search will be available only to the logged-in user; however, the search is saved to the user's account and will appear on whichever system the user logs in to.

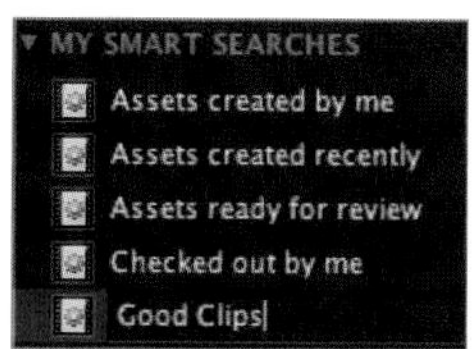

4 Clear the advanced search by clicking the Reset button at the right of the search text field.

TIP To delete a smart search, right-click the entry and choose Delete from the shortcut menu.

Advanced Search Options

The advanced search options allow you to refine your search even further, whether you're searching by keywords or specific metadata fields. You can search any or all fields that are shown in the advanced search options, in addition to keyword searches in the main search field.

For example, you can search for assets checked out by a particular user by using the Checked Out By field. Narrowing down search results to just video clips is accomplished using the Metadata Set field.

Additionally, pop-up menus to the right of each field name allow you to refine the search. Pop-up menu options vary by field, but two major ones are for numbers and text.

The following modifiers are available for text-based fields, such as Title, Location, Stored On, or Annotation:

- All—This field will be ignored.
- Equals—Matches text exactly.
- Not Equals—Matches anything but the text.
- Not Equal and Not Blank—Matches anything but the text only when the field has data in it.
- Contains—Contains the text.
- Begins with—Begins with text in the field.
- Ends with—Ends with text in the field.
- Matches Word—Matches text exactly, using words separated by spaces to determine matches rather than other characters.
- Any Of—Similar to an "or" search, returns at least one word used in the search.

The following modifiers are available for numerical fields, such as Asset ID. Search terms in each of these need to be separated by commas or spaces:

- All—This field will be ignored.
- = —Equal to.
- < —Less than.
- > —Greater than.
- <= —Less than or equal to.
- >= —Greater than or equal to.
- != —Does not equal.

NOTE ▶ To search Final Cut Pro Log & Capture metadata, such as log notes and source timecode from Final Cut Pro projects, activate the advanced search option and set the Metadata Filtering pop-up menu to Include Log & Capture Metadata or Only Include Log & Capture Metadata.

USER ▼

Deleting Assets

Deleting assets in Final Cut Server gives you the option to remove an asset and its primary representation or just the asset from the Final Cut Server catalog.

Therefore, deleting assets from Final Cut Server should be done with great caution, especially if they have not been archived first. Otherwise, the original file will need to be uploaded again.

NOTE ▶ You'll learn about archiving in a later lesson.

There are two methods of deleting assets:

- Select the asset or group of assets and press Command-Delete.
- Choose Delete from the asset's shortcut menu.

In either case, a warning dialog confirms your action before the asset is deleted.

The Remove Asset option removes all traces of asset from the catalog, leaving the original file intact. The file may be uploaded again if desired. "Delete File and Asset" not only removes all representations of the asset from the catalog, but it also deletes the original file from the device—proceed with caution.

NOTE ▶ Only users who are given privileges may delete files.

USER ▼

Locking and Unlocking Assets

Assets can be locked so that others cannot delete or overwrite the asset or its primary representation. There are two methods for doing this:

- Choose Lock from the asset's shortcut menu. You can access the asset's shortcut menu by right-clicking its icon in Thumbnails view.
- Select the asset in a search-results list and click the Lock button in the Toolbar in the main client application window.

A lock icon appears on the asset (in any view) to indicate that you locked it.

Only you or an administrator can unlock an asset that you originally locked. There are four methods for doing this:

- Click the lock icon directly on the asset's icon in Thumbnails view.
- Choose Unlock from the asset's shortcut menu.
- Select the asset in a search results list, and click the Unlock button in the Toolbar in the main client application window.

- Administrators can clear the lock from the Locks pane in the asset info window.

To see who has locked an asset, click the Locks tab in the asset info window.

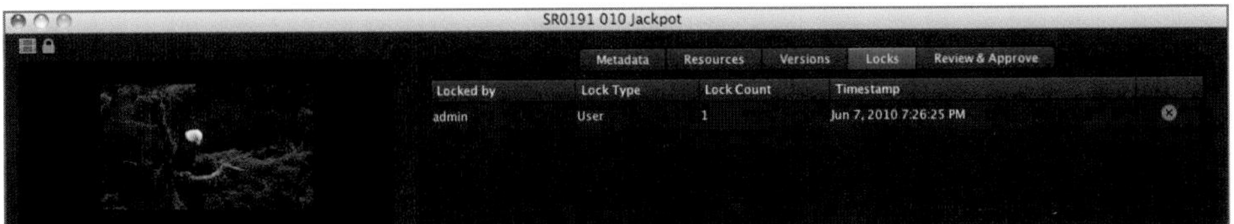

TIP If you encounter a locked asset that you need to work with, you can still carry out many tasks with it, including exporting and duplicating.

USER ▼

Organizing Assets into Productions

Searches provide a fast way to locate your assets, and smart searches are great for one-click recall of those searches. However, there are times when you'll want to group assets together in ways that searches just don't cut it.

Productions allow you to file assets into virtual folders not only for your organizational purposes, but also for sharing with other users, similar to the global smart searches discussed earlier.

There are multiple ways to create productions. They may be created by a user during upload or while a user is perusing the catalog.

Administrators may create productions from files being ingested by scanners or watchers. You'll learn more about these administrator options later.

Let's begin by organizing the Ducati assets via drag and drop.

1 In the client application, click Productions to view the Productions pane.

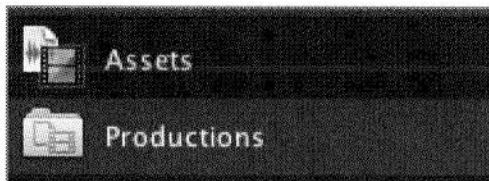

2 To create a new productions folder, click the New Production button.

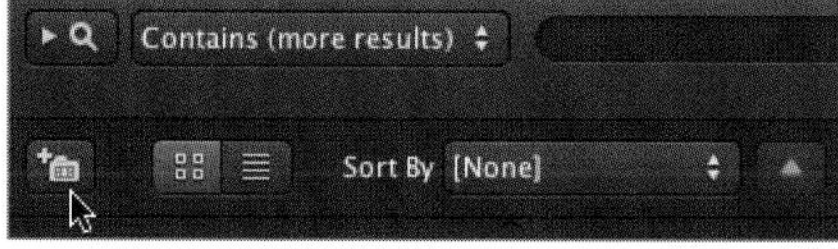

3 In the Production info window that appears, choose Show from the Metadata Set pop-up menu.

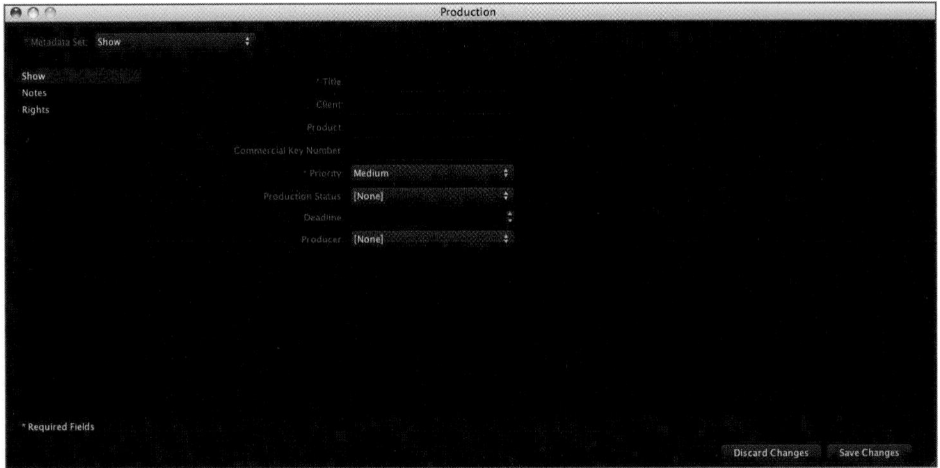

4 Enter some metadata, as shown in the following image, and click Save Changes.

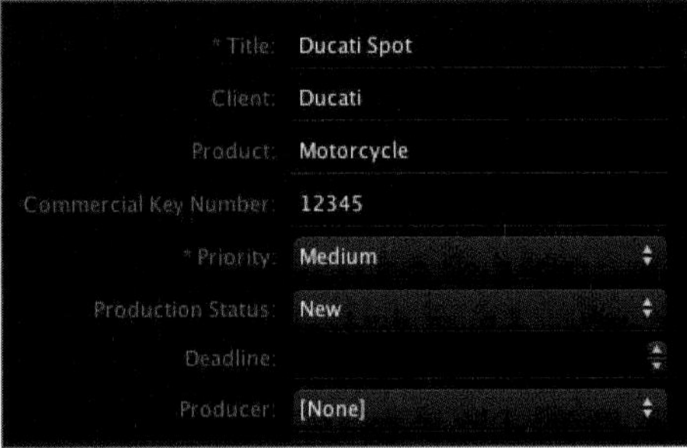

5 Click the Refresh button to update the Productions pane.

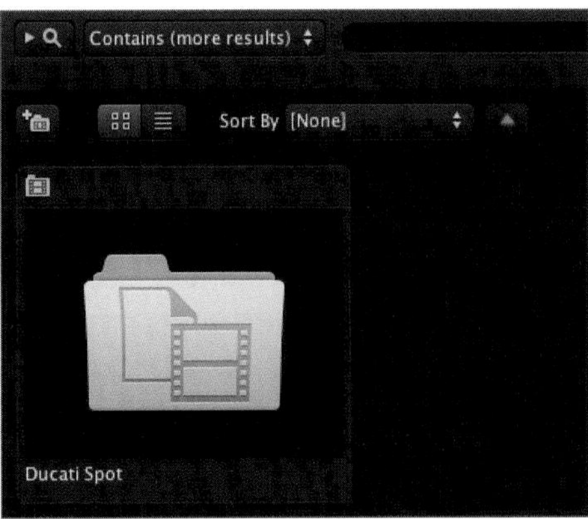

Now that you have created a production folder, you need to populate it with some assets.

6 From the Window pop-up menu, choose New Workspace.

7 In the new workspace, choose the Productions pane. Arrange your windows so you can see both.

8 Drag the **Ducati Managed Project** and **ducati1** assets onto the Ducati Spot production folder.

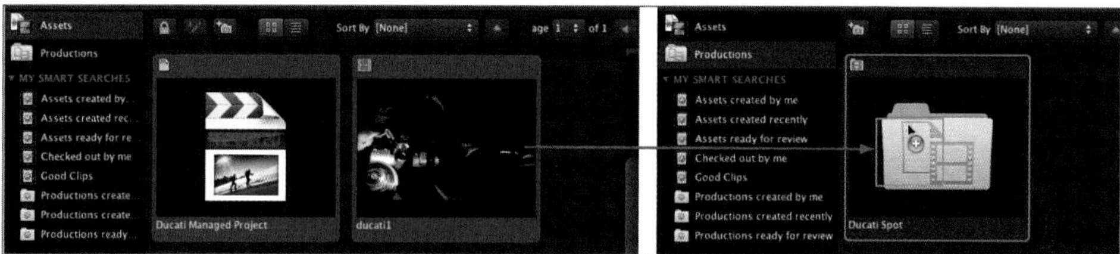

9 Double-click the Ducati Spot production folder to reveal the folder's info window.

The production's info window displays all of the assets linked with this production. If you wish to remove an asset from a production, right-click the displayed asset and choose "Remove from production."

10 Close the Ducati Spot info window, and then close the second workspace window.

Another way to create productions is from a selected group of assets.

11 Back in the Assets pane, perform a search for all assets containing the word *Evergreen*. You should get eight items in the returned results.

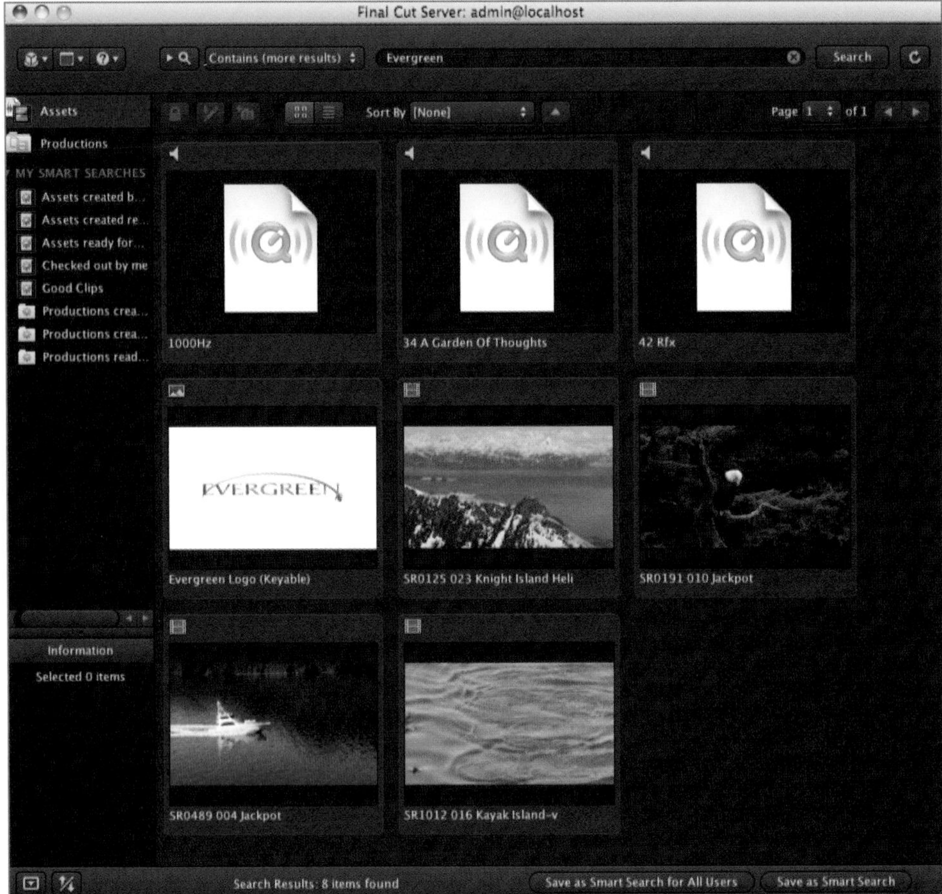

TIP More assets should be associated with the Evergreen project, but only these eight were tagged with *Evergreen* when ingested. The others were not specifically given *Evergreen* as metadata—in some cases because the assets were picked up during an automation.

12 Select one of the assets, and then press Command-A to select all.

13 Right-click one of the selected assets and choose "New Production from Selection" from the shortcut menu.

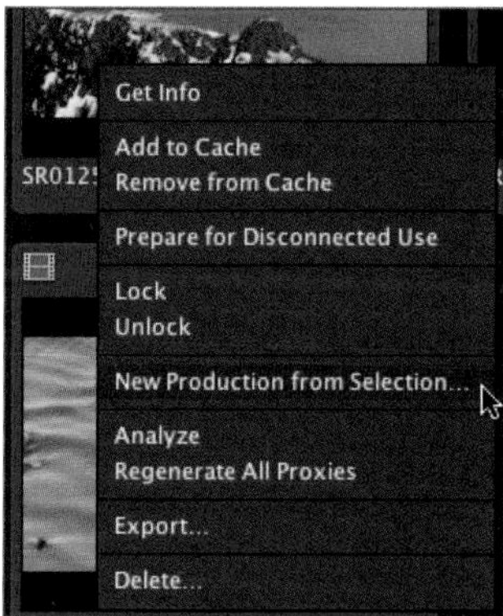

14 In the production info window that appears, choose Show from the Metadata Set pop-up menu and enter *Evergreen* in the Title field. Click Save Changes.

15 Click the Productions pane button, and then click Refresh.

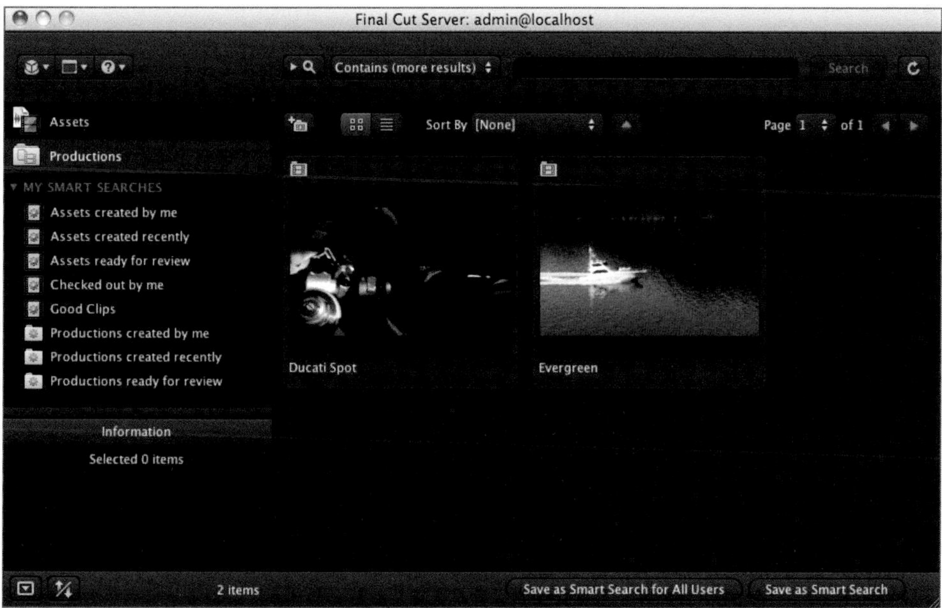

16 Double-click the Evergreen production thumbnail.

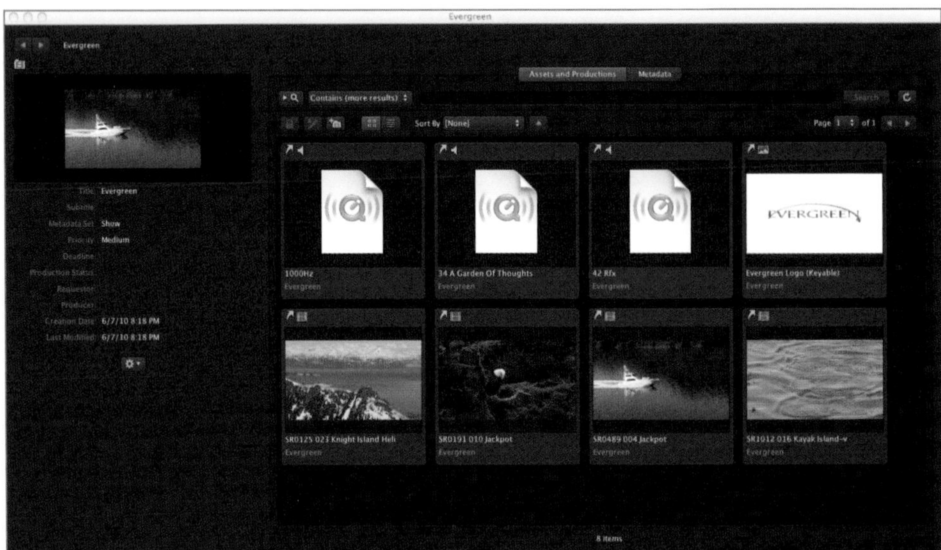

Final Cut Server allows you to create productions within productions. For example, you can create episodic folders within a show's production folder. Or you can create folders to collect similar asset types.

17 Click the New Production folder.

18 From the Metadata Set pop-up menu, choose Show and title the production *Evergreen Audio*. Click Save Changes.

TIP Don't forget to refresh the info window to see the new production folder.

19 Drop the three audio files into the Evergreen Audio folder.

TIP You can double-click a subfolder to see its contents. Use the back arrow (upper-left corner of the window) to navigate back to the parent folder. Final Cut Server allows folders inside subfolders, which means you can micromanage the folders to your heart's content.

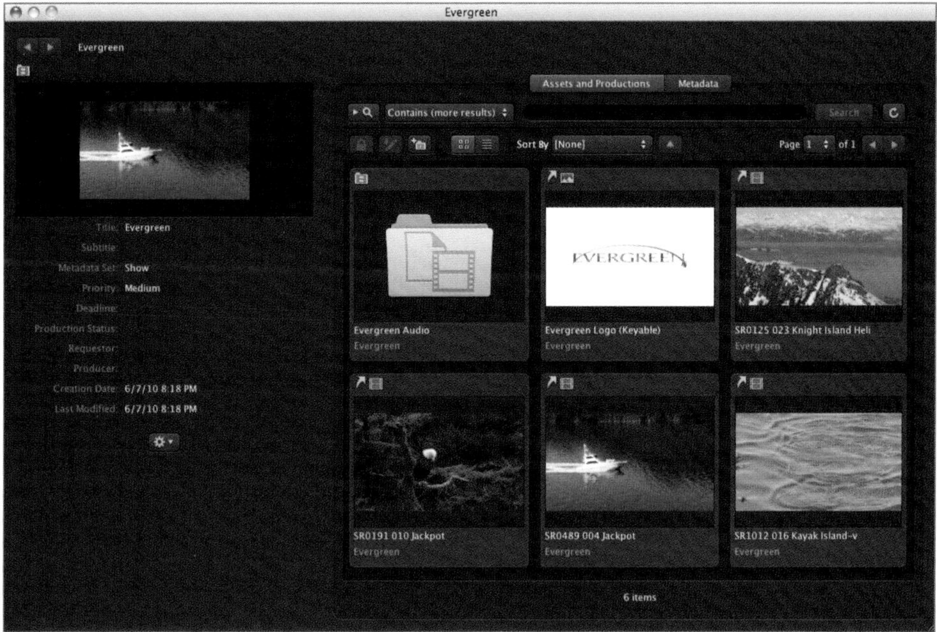

Lesson Review

1. True or false: The search field is case sensitive.
2. In the client application, where can you change the number of search results per page?
3. Which metadata group can you use to modify the metadata fields displayed in the Advanced Search panel?
4. True or false: Any user may delete an asset.

Answers

1. False. Case sensitivity is ignored in searches.
2. You can change the number of search results per page in Preferences.
3. The Asset Filter metadata group can be configured for customized advanced searches.
4. False. A user must belong to a group with delete privileges.

Goals

- Work with Final Cut Pro projects
- Work with online media
- Work with offline proxy media
- Export content from Final Cut Server to Final Cut Pro
- Add content to Final Cut Pro
- Work with version control

Lesson **6**

Editing with Final Cut Pro

In Lesson 3 you learned how to discover and catalog Final Cut Pro assets. This lesson will take you through the process of using these assets in the editorial workflow.

First let's look at some of the key terms used in this lesson:

- Lesson 1 discussed the *primary representation*, the original file that was used to make an asset. In the case of video assets, it can also be referred to as the high-resolution, or hi-res, media. The primary representation will be used for the *online* workflow.
- Some organizations also have *Edit Proxies*, which can use the Apple ProRes Proxy codec. Edit Proxies can be used as a lightweight alternative to the primary representation. Also, in the case where a Final Cut Pro project contains primary representations of different formats, the ProRes Proxy codec serves as an equalizer for all these assets, making it perfect for an editing base. Edit Proxies are used for an *offline* workflow.
- A *edit-in-place* device is one in which the server and clients have access to the media in the same path. Typically, this device will be an Xsan where the server's desktop clients have the Xsan file system mounted at the same path. When you set up an Xsan device in

Final Cut Server it will automatically be set up as an edit-in-place device. It's also possible to configure other devices for edit-in-place use by adding the uniform resource identifier (URI) to the device in the client application's Administration window.

You added a Final Cut Pro project to Final Cut Server in Lesson 3. Let's look in more detail at what happens when the Final Cut Pro project is uploaded to Final Cut Server:

- The Final Cut Pro project file is parsed into an XML file, which is then analyzed for all the media files contained within it. All these media files are linked to the Final Cut Pro project asset and become elements contained within it.
- If a discovered media file is already an asset in the Final Cut Server catalog, the element in the project is simply linked to it, and a new asset is not created.

  If no asset exists in the catalog for any element within the project file, those media files are automatically uploaded to the same device and path as the Final Cut Pro project file. Further, assets are automatically made for those media files, and links are established between the new assets and the elements within the Final Cut Pro project asset. The exception to this is if the media files reside on an edit-in-place device (such as an Xsan). In that case, the media files will not be moved but will still have assets and elements created for them.

ADMIN ▼

Preparing Final Cut Server

To prepare for this lesson, you will check some configuration options in Final Cut Server. Then, you'll create two devices. One of these will be a standard local device, where media needs to be cached locally before it can be used. The other will simulate working in an Xsan environment where you have direct access to the media.

Checking Edit Proxies' Configuration

Edit Proxies can be configured at the time of installation. By default, the setting is off. Here, you will enable Edit Proxies, and configure them to be created using the ProRes Proxy codec.

1. In the client application, open the Administration window, and go to the Preferences pane.

2 Click the Proxies tab. Select the Enable Edit Proxies option if necessary.

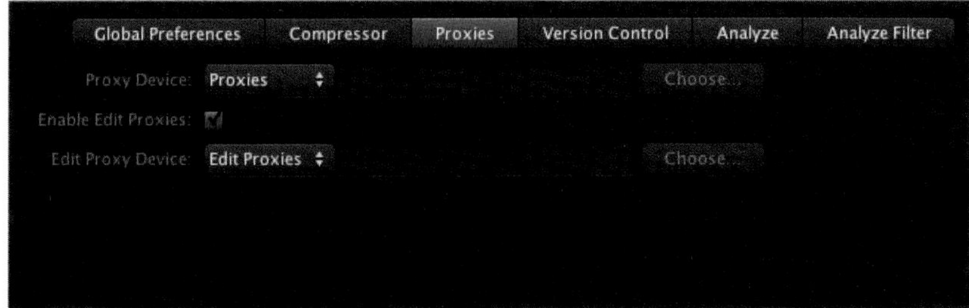

3 Click the Analyze tab. Here you are able to select what transcode settings are used for various proxy types. For Clip Edit Proxy, set the Transcode Setting to Apple ProRes 422 (Proxy) Clip Edit Proxy. Click Save Changes.

Checking Versioning Configuration

Versioning can also be configured at the time of installation. By default, the setting is on. You can change the setting after installation from the Administration window.

When versioning is enabled, it's triggered for assets that are checked in. The new file replaces the file on the device being uploaded to. The replaced file is moved into the Versions device.

Later in the lesson, you'll learn where you can see the different versions of an asset, and how you can either revert to the original or make a new asset from the version you see.

Now you'll check to see that versioning is enabled.

1 In the Preferences pane, click the Version Control tab, and ensure that the Create Asset Version Default option is selected.

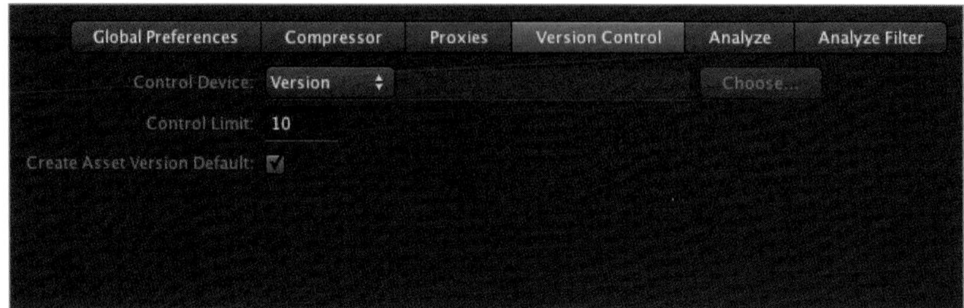

The Control Limit specifies the number of versions Final Cut Server will keep for an asset. When that number is reached, the oldest version gets removed at each subsequent checkin.

2 Close the Administration window and log out of the client application.

Creating a Standard Local Device

Now you'll create a standard local device to show what happens when the client does not have direct access to the media. The folder you will use for the device contains some media items that will be scanned by Final Cut Server.

1 In the FCS_Book_Files folder on your desktop, locate the Leverage Local folder.

2 Drag the Leverage Local folder to the Macintosh HD/FCSvr folder.

3 Open System Preferences, select the Final Cut Server pane, and then authenticate.

4 Click the Devices button and click the Add (+) button to add a new device.

5 In the Device Setup Assistant, select Local and click Continue.

6 Enter *Leverage Local* for Device Name and navigate to the Leverage Local folder. Click Continue.

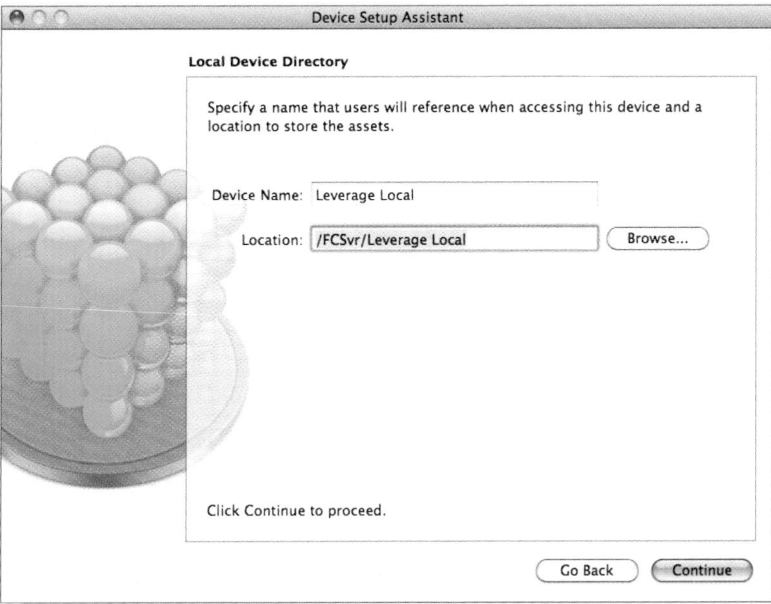

7 Leave "Enable as an Archive Device" deselected and click Continue.

8 Select Full Scan and set the kickoff time to be approximately 4 minutes from your server's current time. Select Add Only Scan and set it to kick off every 5 minutes. Leave Metadata Set at Media and click Continue.

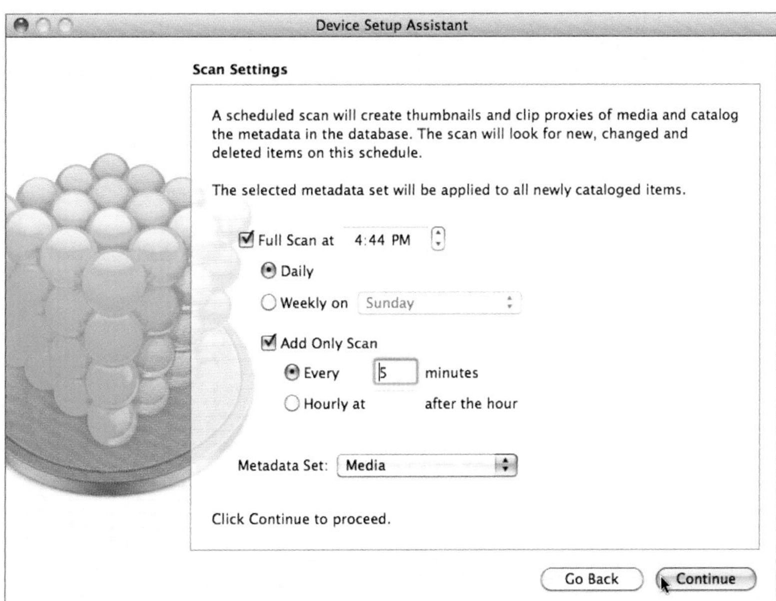

9 In the Transcode Settings screen, leave No Conversion selected. Click Continue.

NOTE ▶ It's unnecessary at this time to select any other codecs, as in this exercise you will not be copying content to this device. If you do select additional codecs here, they'll be available as a conversion option when you copy media to the device.

10 Review the final settings, and make sure they're identical to the following figure (except for time, which will differ). Click Done when you're finished comparing. If you've made any mistakes, you can click Go Back and fix them.

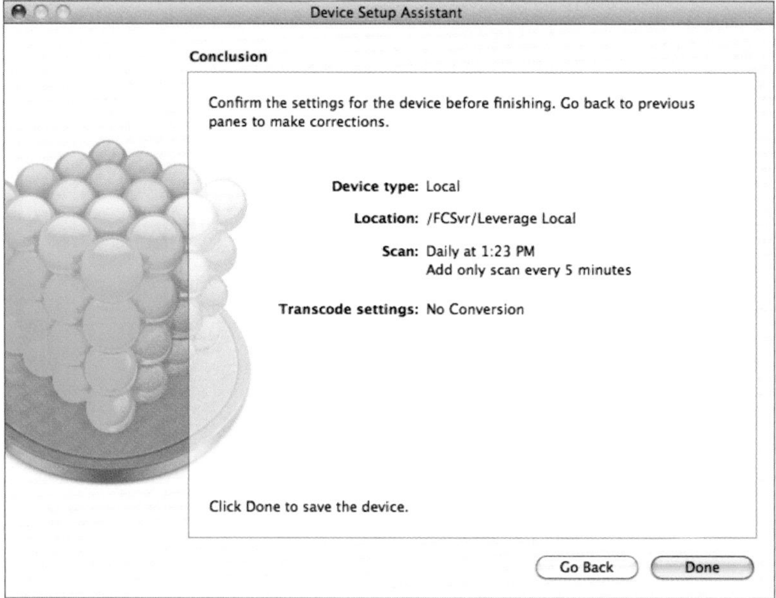

Depending on what time you set your scans to run, within 5 minutes Final Cut Server should start automatically discovering assets from the new device and adding them to the catalog. You can open the Search All Jobs window to watch the proxies get created in real time.

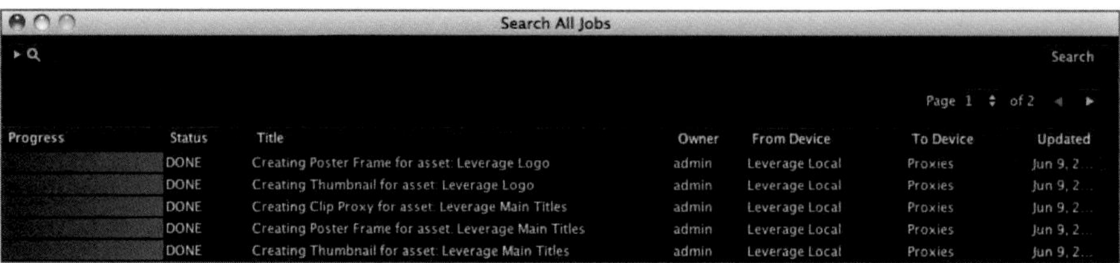

Device setup is now complete. You'll use an asset on this device later in this lesson.

Creating an Edit-in-Place Device

Because you're working on only one machine for these lessons—and are running the client and server on the same machine—you're able to simulate an Xsan environment by turning a local device into an edit-in-place device.

1 Create a folder called *Leverage EIP* in the FCSvr folder at the root of your drive.

2 If necessary, open System Preferences and then the Final Cut Server pane.

3 Click the Devices tab and click the Add (+) button to add a new device.

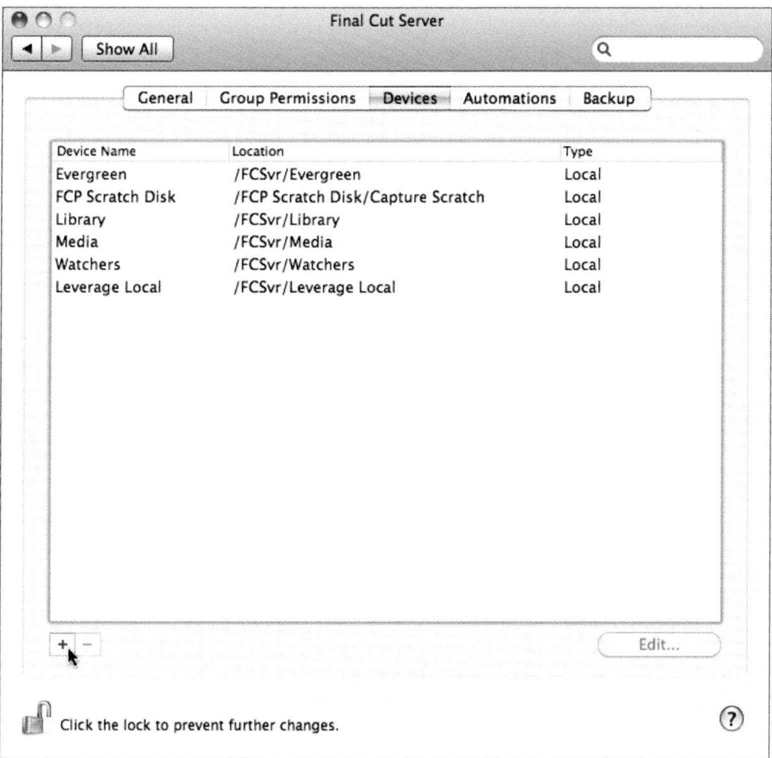

4 In the Device Setup Assistant, select Local and click Continue.

Although you are about to simulate an Xsan, do not choose to add an Xsan device here. The Device Setup Assistant will recognize that no Xsan is available.

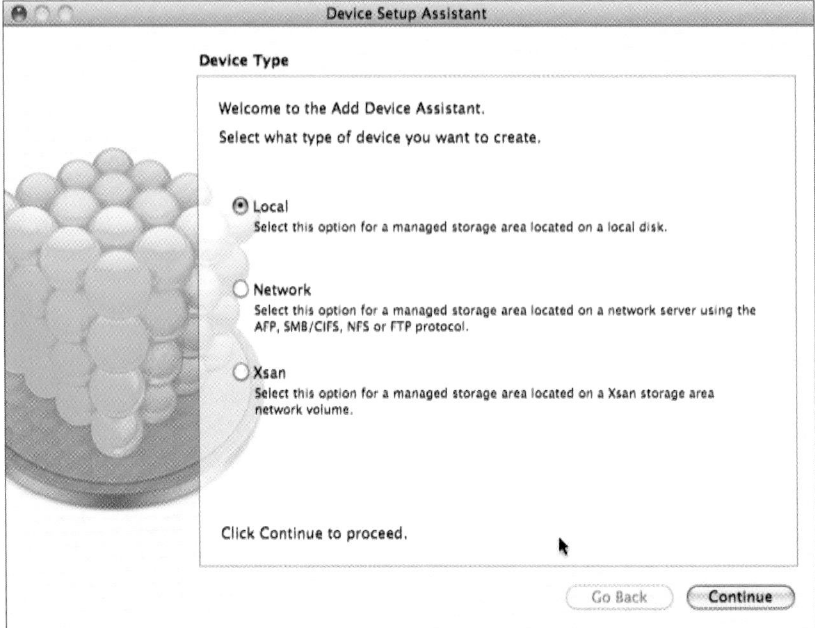

5 Enter *Leverage EIP* for Device Name and navigate to the Leverage EIP folder. Click Continue.

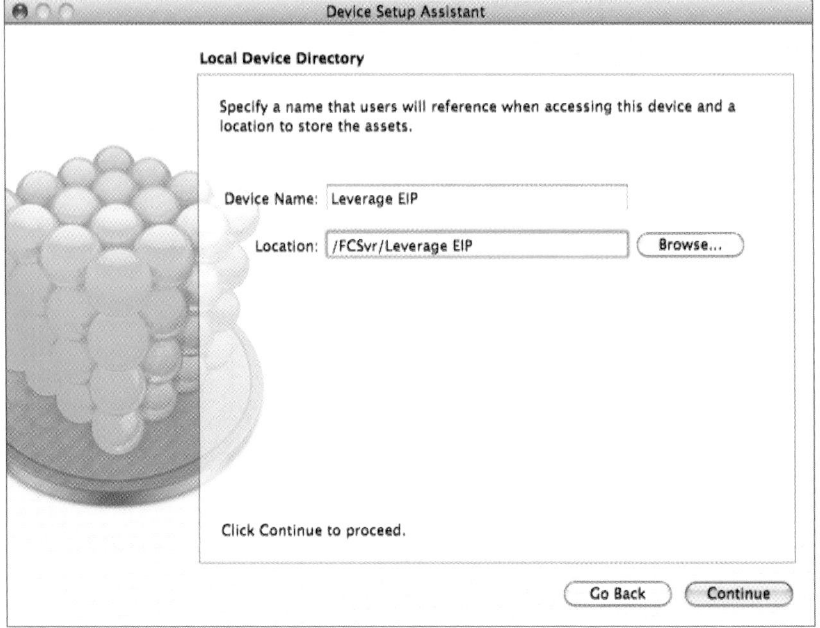

6 In the next pane of the Device Setup Assistant, leave "Enable as an Archive Device" deselected and click Continue.

7 Leave Full Scan deselected and click Continue. In this exercise you're adding content to this device using the client application and you're not scanning.

8 Leave No Conversion selected. It's unnecessary at this time to select any other codecs. In this exercise you're using this device as a source device, not a destination device, which would require a transcode. Click Continue.

9 Review the final settings, and make sure they're identical to those in the following figure. Click Done when you're finished comparing. If you've made any mistakes, you can click Go Back and fix them.

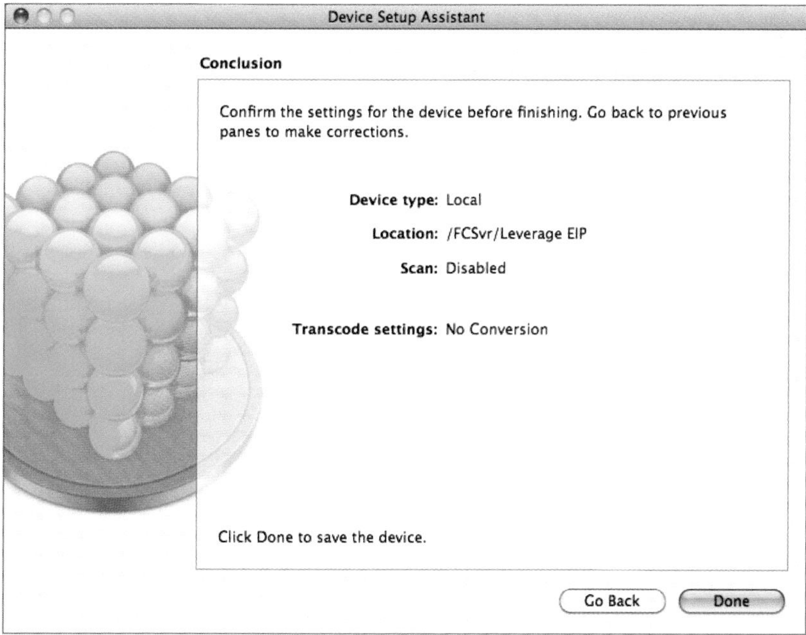

To change this device from a regular local device into an edit-in-place one, you need to modify the device settings in the client application's Administration window.

10 Quit System Preferences.

11 Information about devices is cached in the client when it starts. If the client is open, quit and then start it again.

12 Open the client Administration window and then the Devices pane, and double-click the Leverage EIP device to open it.

13 Copy the path from the Local Directory field to the "Macintosh edit-in-place URI" field.

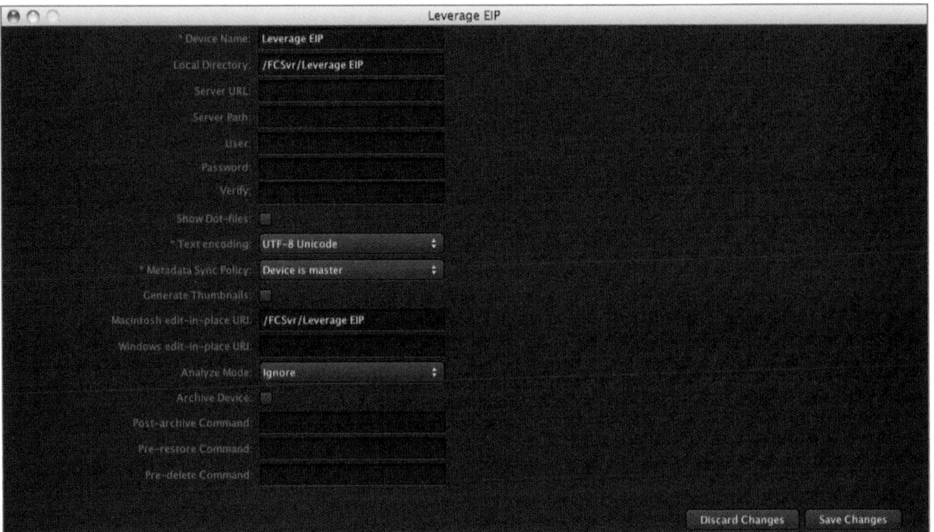

14 Click Save Changes.

When you do this, the path you copied in will automatically convert to a valid URI string—in this case, "file://localhost/FCSvr/Leverage EIP."

15 Device setup is now complete. As in step 11, you will need to quit and reopen the client again to ensure that the updated edit-in-place device information is used.

USER ▼

Adding Final Cut Pro Projects to Final Cut Server

In Lesson 3 you discovered how to add Final Cut Pro projects to Final Cut Server. You're now going to add an existing project to the Leverage EIP device and then use it in an online and an offline workflow.

1 Open the Leverage for Upload folder inside the FCS_Book_Files folder on your desktop. Drop the **Leverage EP 205** Final Cut Pro project file into the client application. The Upload Final Cut Pro Project window opens.

2 Set the Destination to *Leverage EIP*.

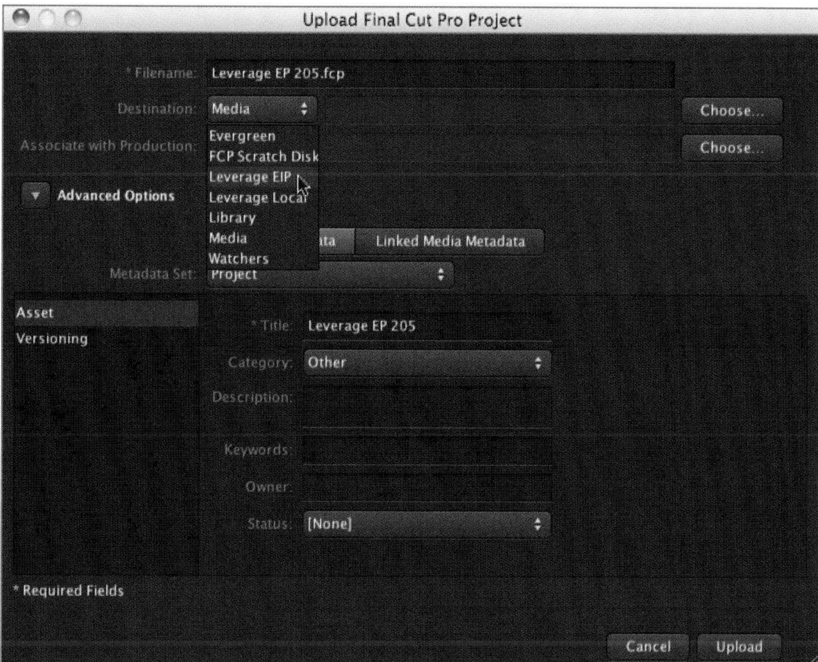

3 Under Project Metadata, choose Project in the Metadata Set pop-up menu.

The Project Metadata Set value determines which metadata will be given to the Final Cut Pro Project asset. Add some descriptive metadata, as shown in the following figure.

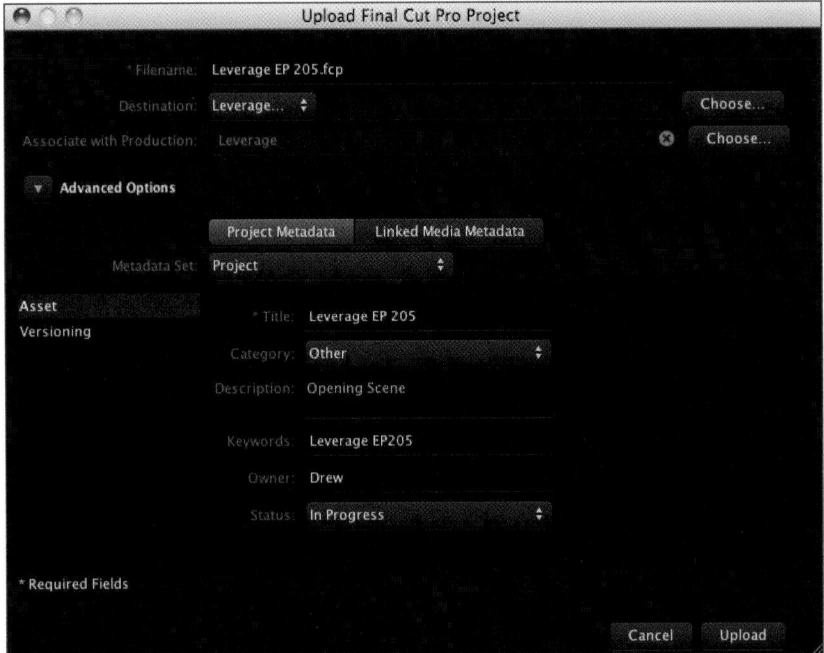

4 Select the Versioning metadata group, and check that the "Store and track versions of this asset" checkbox is selected.

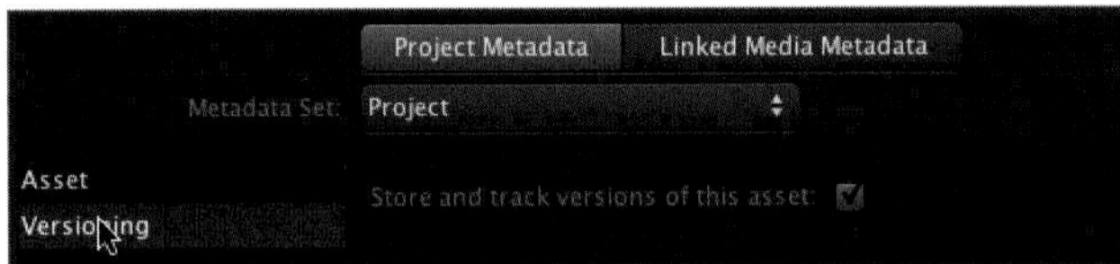

5 Click the Linked Media Metadata tab. This will set the Metadata Set for all the assets contained in the Final Cut Pro project. Add some descriptive metadata.

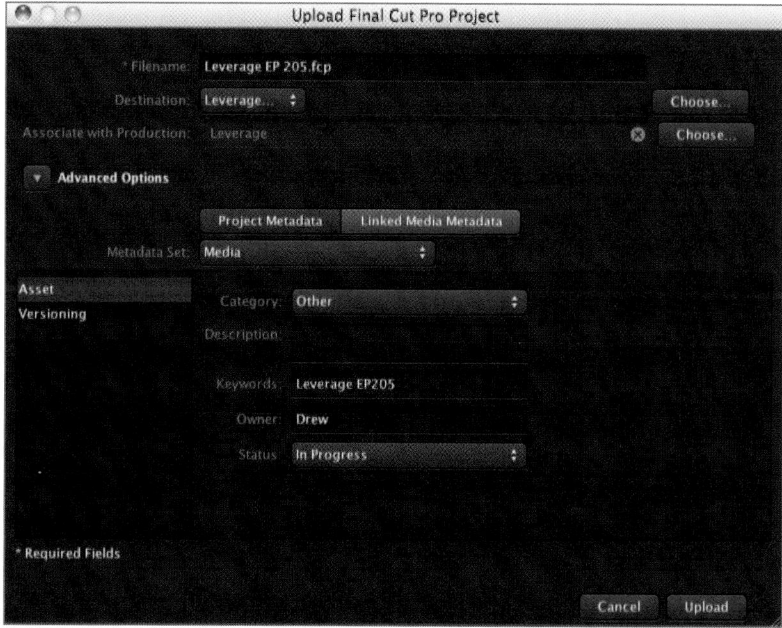

6 Click the Upload button.

Final Cut Server uploads the Final Cut Pro project to the Leverage EIP device and then analyzes it to determine what media files are contained in the project. Once this is done, the media files will be uploaded to the device in a subfolder at the level where the project file was placed. Assets will be created for each of the media files. You can open the Search All Jobs window to watch the proxies get created in real time.

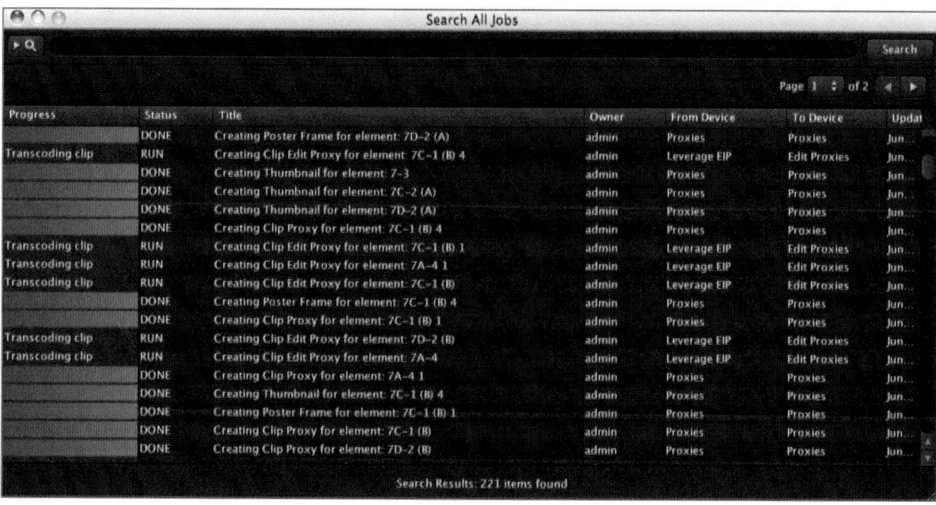

When the jobs are complete, you'll create a production to gather all the Leverage assets together.

7 In the Assets pane, enter *Leverage* in the Search field.

8 Right-click the **Leverage EP 205** Final Cut Pro project asset and choose "New Production from Selection" from the shortcut menu.

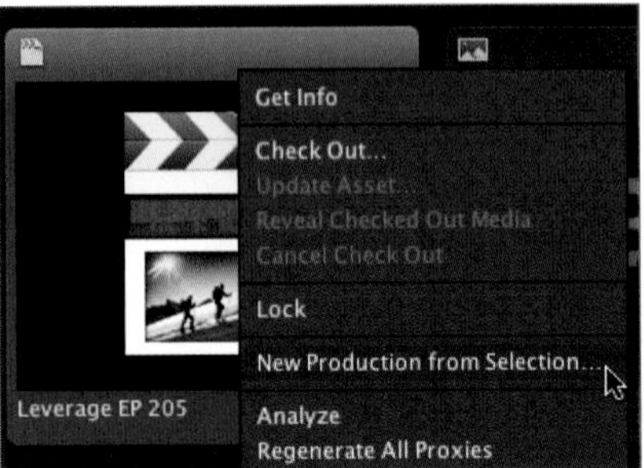

9 Set the Metadata Set of the production to Show. Enter *Leverage* as the Title, and then click Save Changes.

10 In the sidebar, choose the Productions pane, and then click the Refresh button.

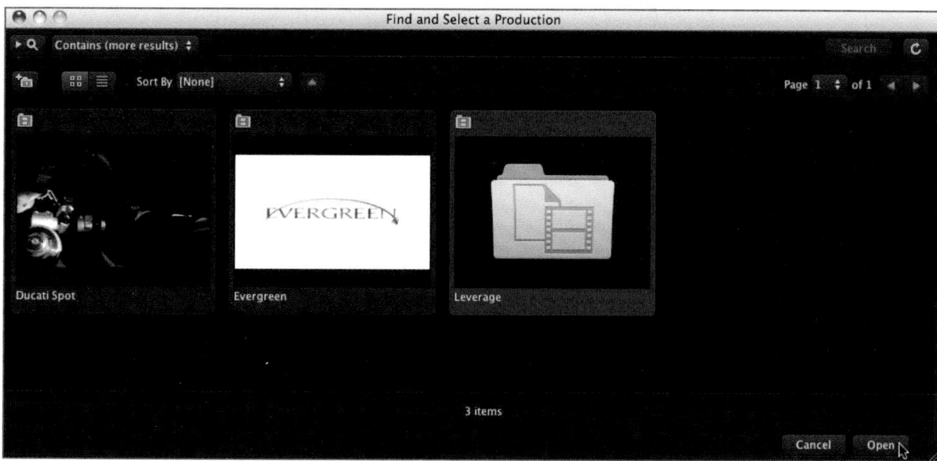

11 Double-click the Leverage production folder to see the project you uploaded.

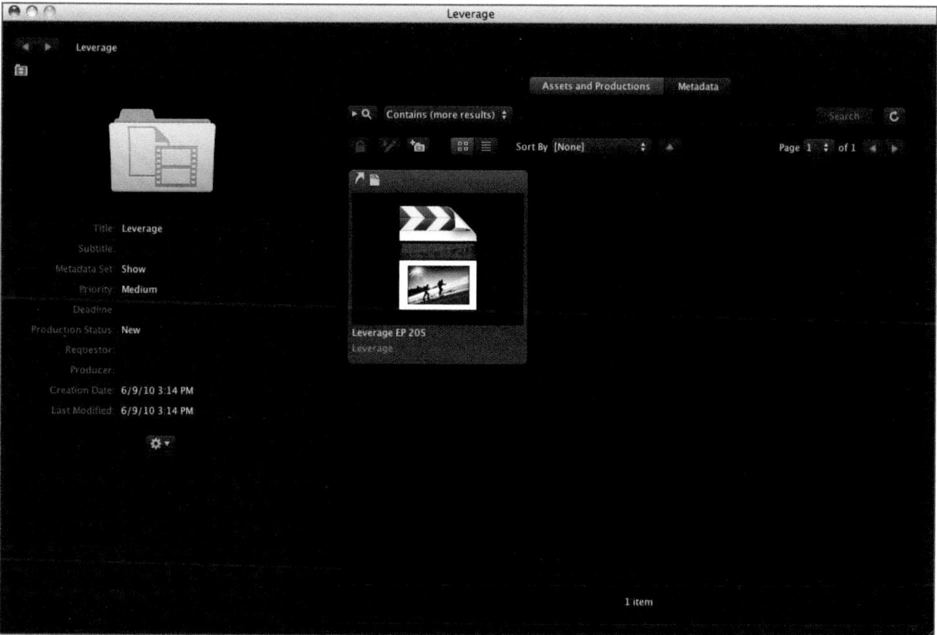

12 Double-click the project to see all the media files contained in the project as elements.

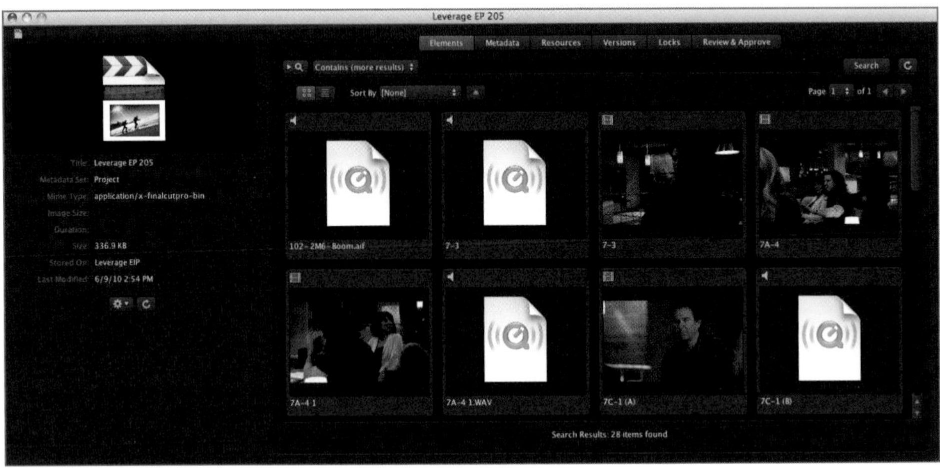

13 Close the asset info window, returning to the Leverage production info window.

You will now see how you can use this project in online and offline workflows.

USER ▼

Online Workflow

In Final Cut Server, an online workflow implies that you are working with the primary representations of the media assets, and that these assets are directly accessible from the machine you're working from. This is the typical usage in an Xsan environment.

Final Cut Pro projects are checked in and out of Final Cut Server. In the online workflow, media cataloged as assets in Final Cut Server will not be moved. Instead, the project will directly link to these files. This means that there's no waiting for files to be copied to the local machine; editors can work on the project as soon as the Final Cut Pro project is checked out.

In the following exercise, you'll learn how to apply an online workflow.

1. Create a folder on the desktop called CHECKOUT, and drag it to Places in the Finder sidebar.

   You can use this single location for Final Cut Server checkouts. After you check out the file, don't move it; Final Cut Server will look for the asset at the location to which you checked it out.

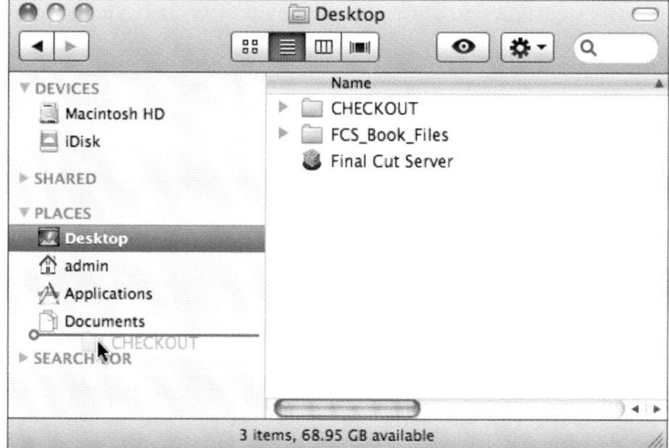

2 Select the Leverage project that was uploaded earlier, right-click it, and choose Check Out from the shortcut menu.

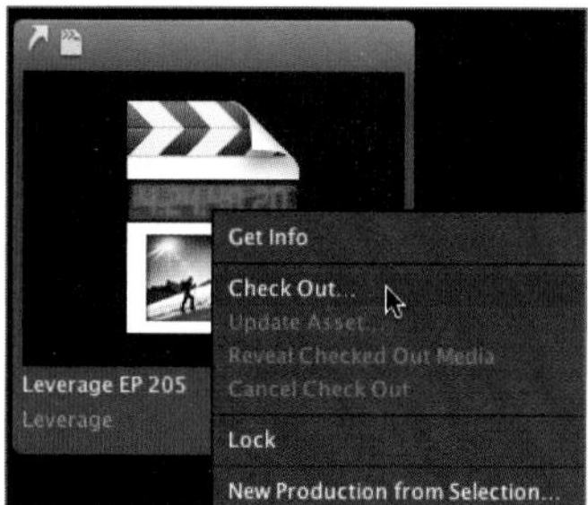

The Check Out window opens.

3 Click Choose to select the location that the project file is to be checked out to. In this case, choose the CHECKOUT folder you created on the Desktop. Click Open.

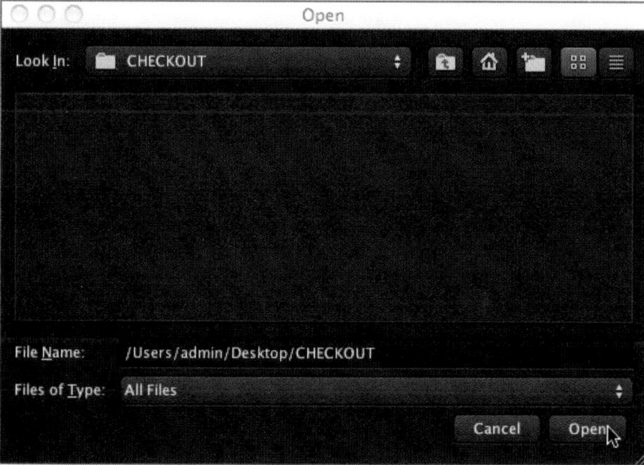

4 Keep the Use pop-up menu set to Original Media, and click Check Out.

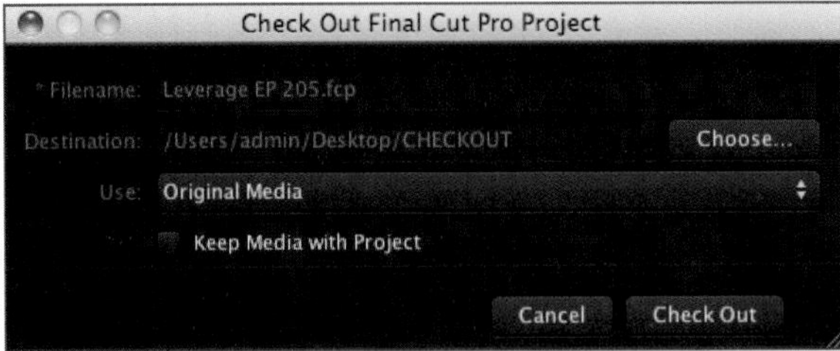

When the checkout finishes, the asset will be updated to checked-out status.

5 Go to the CHECKOUT folder and open the checked-out project file.

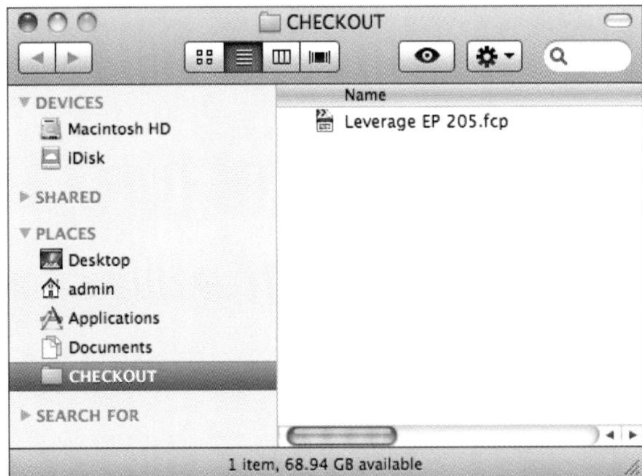

6 Right-click the first clip in the Final Cut Pro sequence and choose Item Properties > Format from the shortcut menu.

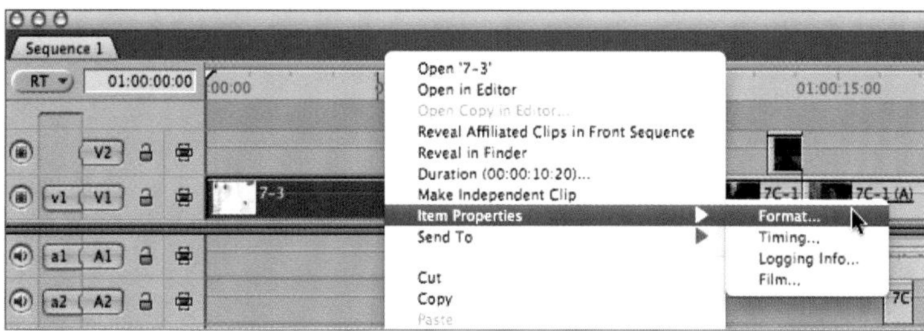

If you look at the Source field, you'll see the clip is still in the same location on the Leverage EIP device.

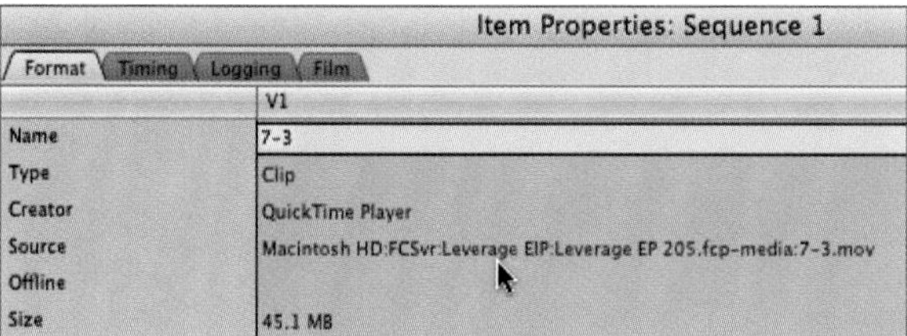

7 Make some simple edits, such as deleting some clips from the sequence, and then save the project. Quit Final Cut Pro.

8 Go back to Final Cut Server, right-click the checked-out project, and choose Check In from the shortcut menu.

9 Enter *Deleted some clips* for Version Comments, and then click Check In.

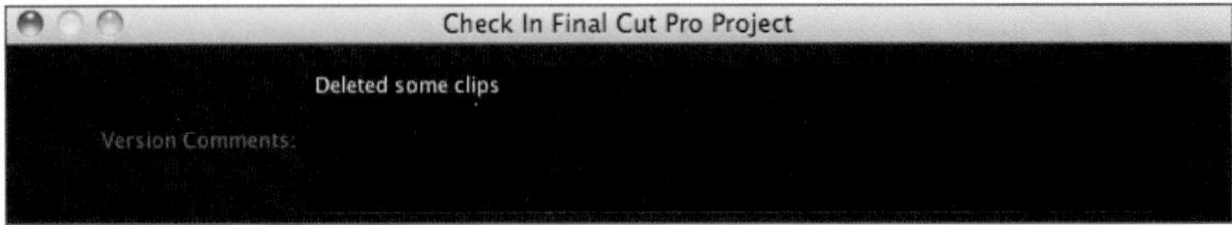

When you check in the project, Final Cut Server reanalyzes the project file. The device where the project file exists will be updated with the new copy, and the old copy will be copied into the Versions system.

No new assets were created as no new media objects had been added to the project while it was checked out.

If any media objects are added to a project while it's checked out, they will be uploaded to the same device where the project exists. Each object will be cataloged as an asset, and each will be added to the Final Cut Pro project asset as an element.

USER ▼

Offline Edit Proxy Workflow

If you're not connected to an Xsan, and the client that is using it doesn't have direct access to the media, you have two ways to work with Final Cut Server. First, you can work with the primary representations. In this scenario, the media files will be copied from the source device and cached on the local machine.

The second option is to use Edit Proxies. The ProRes Proxy format provides some significant size advantages while maintaining excellent quality.

Using Primary Representations and Caching Locally

If you check out a project and it contains media assets that are not on an edit-in-place device, Final Cut Server will automatically cache the files on the client machine. The location of this cache is determined in the Preferences window on the client application.

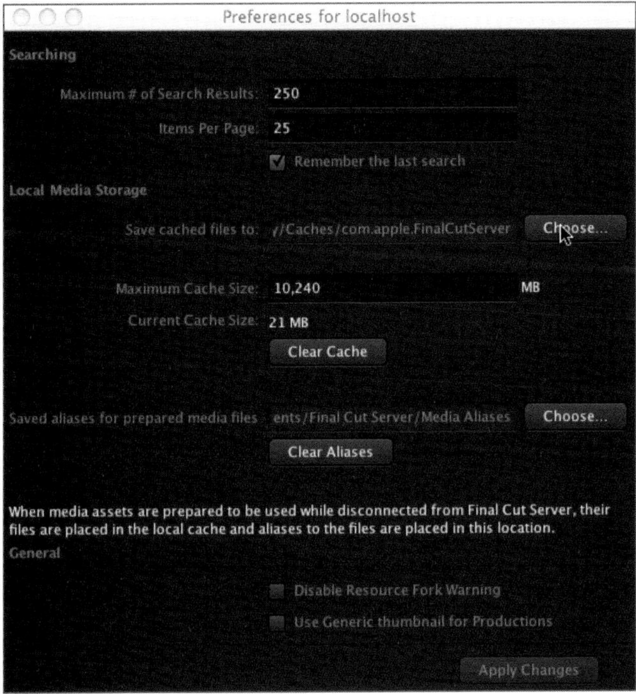

1 Select the Leverage project, and click Check Out. This time, select the "Keep Media with Project" checkbox.

The project will download, and all the media will be copied locally into the CHECKOUT folder.

2 Navigate to the Leverage EP 205 folder inside the CHECKOUT folder.

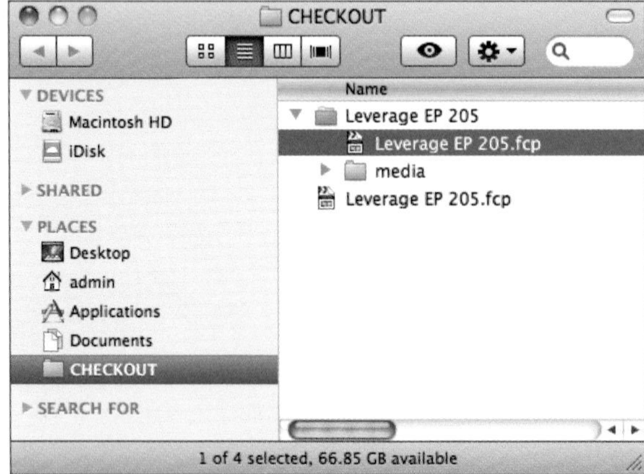

3 Open the **Leverage EP 205.fcp** project to confirm that the media files are referencing this new location.

4 Right-click the first clip in the sequence and choose Item Properties > Format from the shortcut menu.

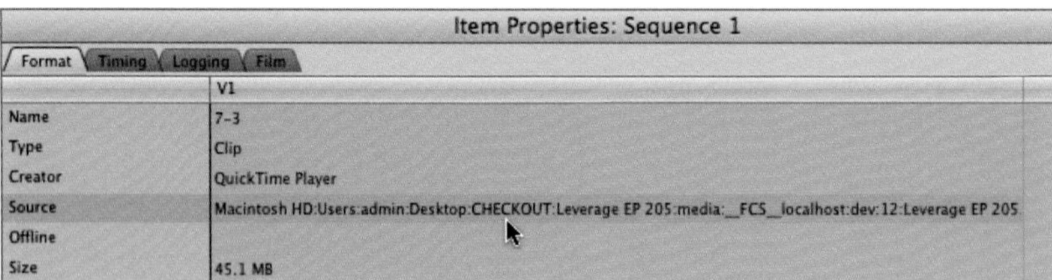

Note that the source media file for the clip is referencing the media files inside the CHECKOUT folder.

5 Click OK, save the project, and then quit Final Cut Pro.

Cancel Checkout

You may find at some point that you accidently checked out the wrong project, or as in this exercise, you need to check out a Final Cut Pro project with Edit Proxies. You can cancel a checkout without having to check the project back in.

1 In the client application, right-click the **Leverage EP 205** Final Cut Pro project file.

2 Choose Cancel Check Out from the shortcut menu.

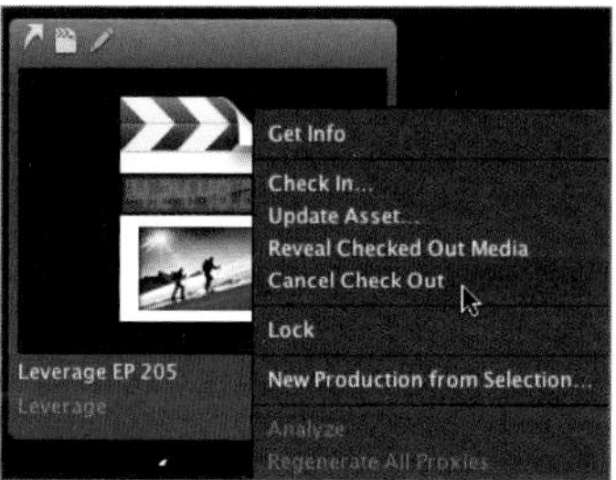

Canceling checkout does not remove the copied assets. To recover some storage space and to prepare for the next exercise, you will delete the current items in the CHECKOUT folder.

3 In the Finder, delete all the contents from the CHECKOUT folder on your desktop.

Using Edit Proxies

For all the assets contained in a Final Cut Pro project, Final Cut Server will create ProRes Proxy Edit Proxies. These are great for editing in an offline scenario, or on a laptop, as the file sizes will be significantly smaller than the primary representations. However, the raster size is maintained, and they offer excellent quality for an offline workflow.

1 Back in the client application, select the **Leverage EP 205** project file and check it out again.

2 This time, in the Check Out window, choose Edit Proxy in the Use pop-up menu, and then click Check Out.

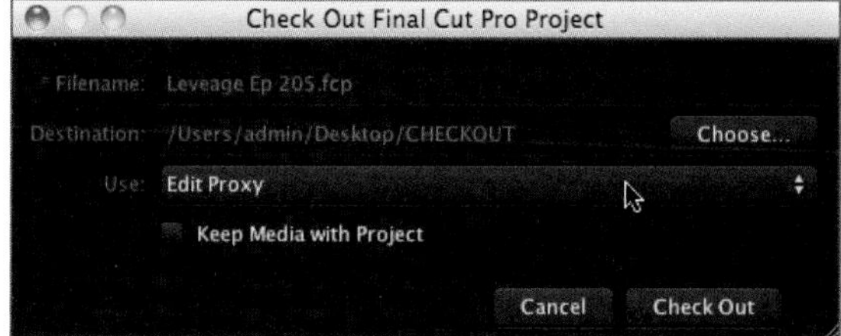

3 Not every asset type has an Edit Proxy. If such assets are in your project, a dialog opens, letting you know that some assets do not have Edit Proxies. Click Continue.

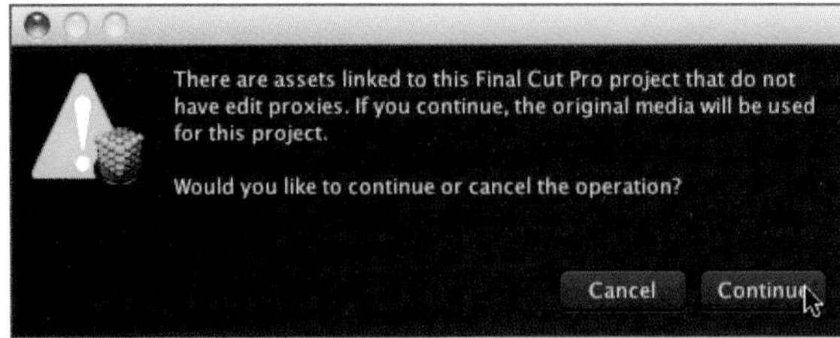

Once the project is checked out and the Edit Proxies are cached on the local machine, you can now open the project in Final Cut Pro.

Any changes made to the project while you're using the Edit Proxies will be reflected in the project when you check it back in.

The project could now be checked out using the primary representations. An editor finishing the project for export, or doing other work requiring the primary representations, such as color correction or motion graphics, could do this.

Adding New Content to a Final Cut Pro Project

You can add media to a Final Cut Pro project in two ways: either by importing the media direct from Final Cut Server, or by adding it to the Final Cut Pro project and then checking the project into Final Cut Server. The next two exercises describe how each approach is used.

Dragging from Final Cut Server

When working in Final Cut Pro, you can use Final Cut Server to search for content that you can then drag to your Final Cut Pro project. There are two scenarios, depending on whether the media is on an edit-in-place device and directly accessible (for example, in an Xsan environment), or whether the media needs to be cached. You'll look at the each scenario in turn.

Adding New Content from an Edit-in-Place Device

In an Xsan environment, you may have direct access to the media that you are adding to a Final Cut Pro project. To add media to an open Final Cut Pro project, follow these steps:

1 If you canceled the last checkout, check out the Leverage project again. Open the checked-out project in Final Cut Pro.

2 Open the Final Cut Server client application.

3 In the Assets pane, search for *Leverage*. Scroll down to the last 8 to 10 assets.

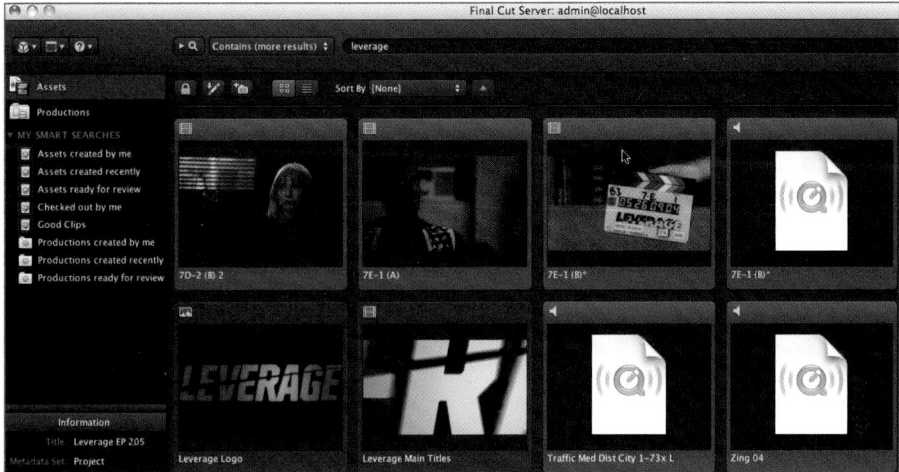

Note that the frames of some assets are a lighter gray than some others, which means that they can be used immediately without caching. In this case, you're seeing the lighter-gray frames because the asset is on your simulated Xsan edit-in-place device, Leverage EIP.

4 Arrange the windows as necessary so you can now drag a lighter-gray asset from Final Cut Server directly to your Final Cut Pro Browser or Timeline.

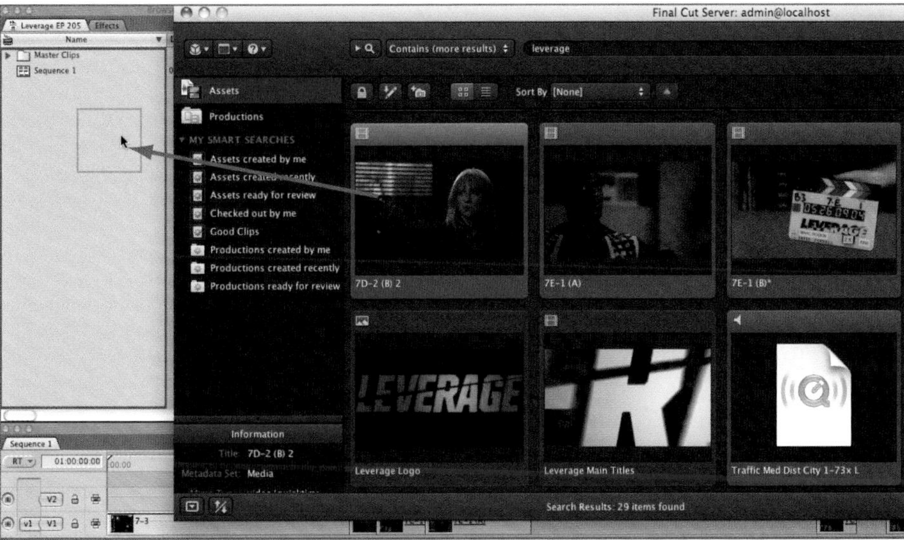

When you next check in the Final Cut Pro project to Final Cut Server, the additional assets will be added to the Final Cut Pro asset as elements.

NOTE ▶ You can add local media to your Final Cut Pro project while it's checked out using either Edit Proxies or primary representations. Any media that is added will be uploaded to Final Cut Server and cataloged as assets when the Final Cut Pro project file is checked in.

Adding New Content from a Non-Edit-in-Place Device

The key difference between adding media from edit-in-place devices and media that is not on an edit-in-place device is that media from non-edit-in-place devices must first be cached locally before it can be used.

The Leverage Local device you created earlier contains some media assets. You will add a clip from it to your project.

1 In the client application, locate the **Leverage Logo** asset.

Note that the asset frame is a dark gray. This means that it needs to be cached locally before it can be used. There are two ways to do this: either by dragging the asset over an application (such as Final Cut Pro), or by Control-clicking (Mac) or right-clicking (Windows) the asset and choosing Add to Cache from the shortcut menu.

2 In this case, drag the asset into the Final Cut Pro Browser. A dialog will appear. Click Add to Cache.

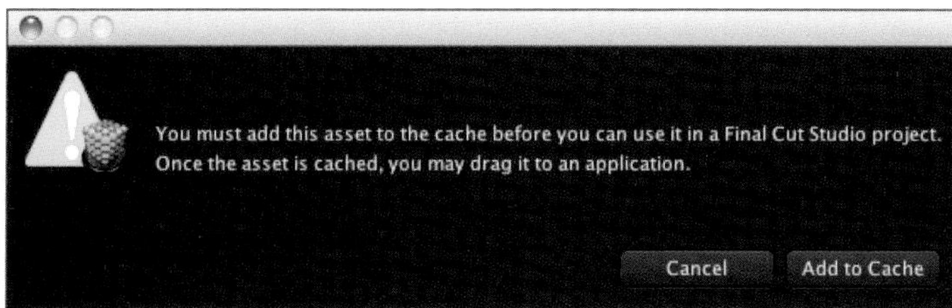

The asset's frame will switch to light gray once the asset is cached.

3 Drag the asset to the Final Cut Pro Browser.

4 Drag the clip from the Browser to the Timeline of the project. Save the project and quit Final Cut Pro.

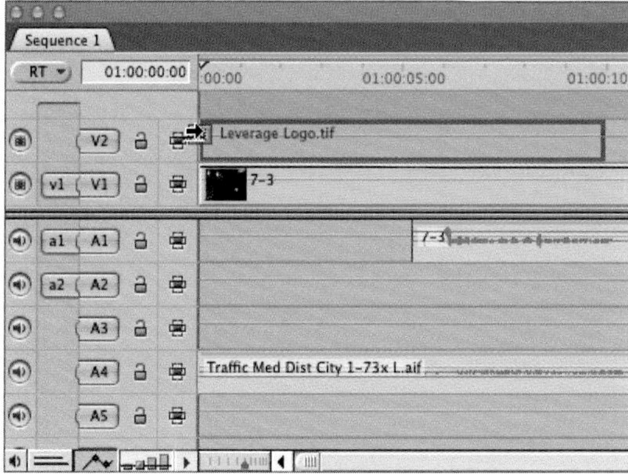

5 Go back to the Leverage productions folder and check the project back in.

6 Enter *Added graphic* into the Version Comments field, and then click Check In.

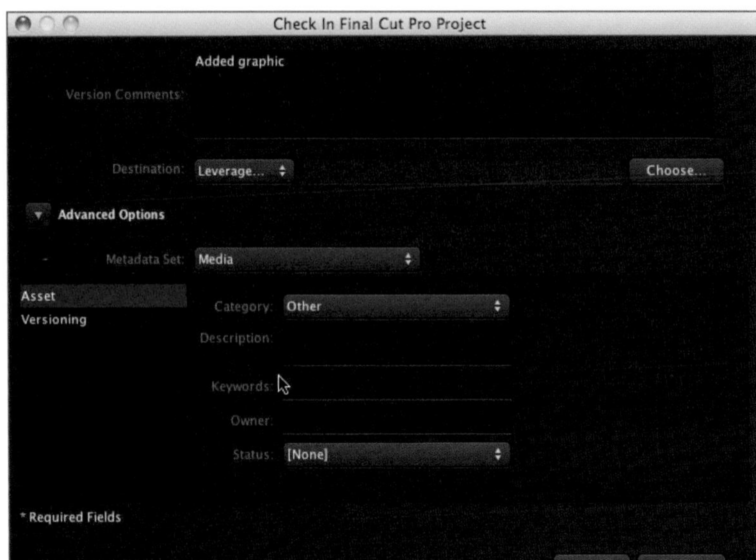

7 Once the check-in is complete, double-click the project asset and go to the Elements tab. Note the **Leverage Logo.tif** asset has been added to the project as an element.

Using Version Control

When version control is enabled, it will be activated on all uploaded assets. For each asset that is checked out and then checked back in, you will see a Version Comments field where you can make entries, as you have done a couple of times in this lesson.

For a regular asset, you will see the following window when checking back in:

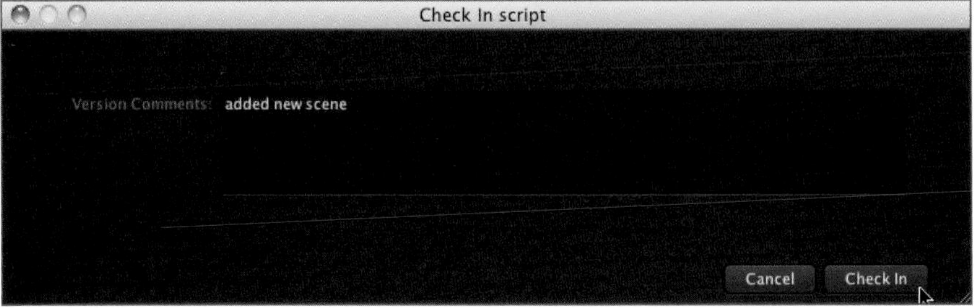

In the case of a Final Cut Pro project asset, you'll see a larger window:

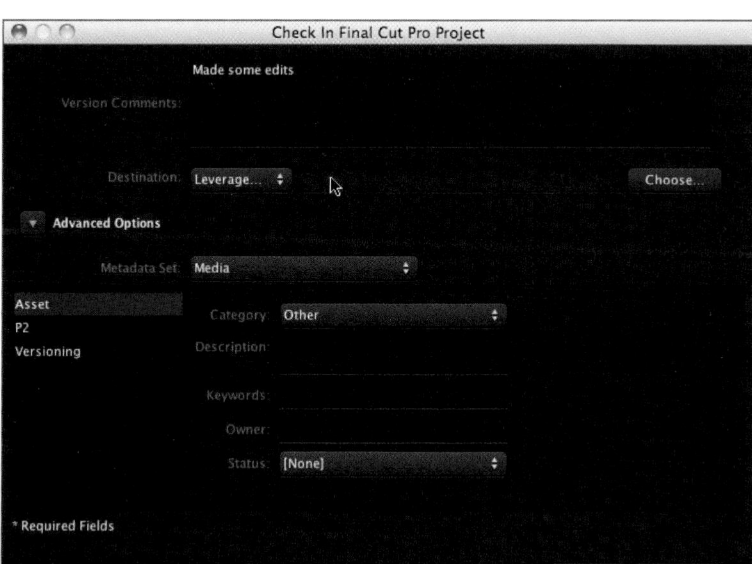

The Destination and Metadata information in this window have been applied to any new media files that have been added to the Final Cut Pro project while it has been checked out.

The different versions can be seen in the Versions pane in the asset info window. Now you'll enable version control and see what it allows you to do.

1 In the **Leverage EP 205** asset info window, click the Versions pane.

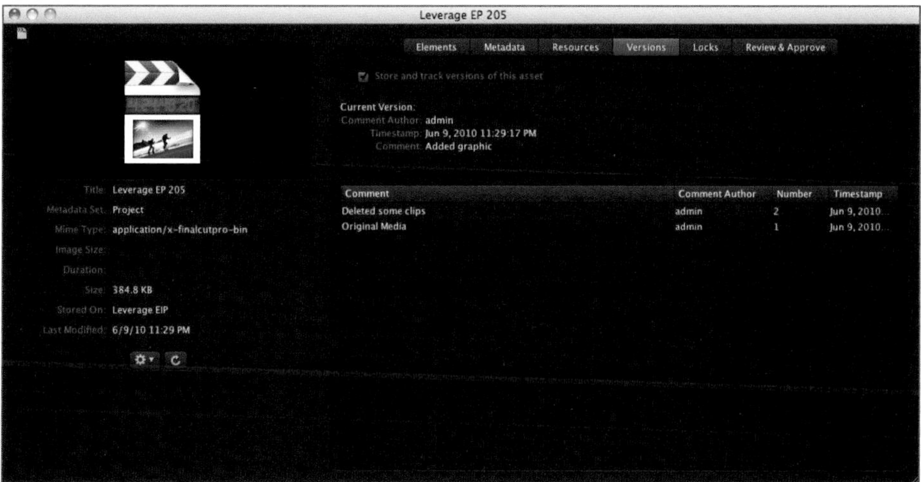

Here you can see the comment author, version comments, and the version number.

If you right-click one of the versions, you can choose to revert to it, view it, or create a new asset from it.

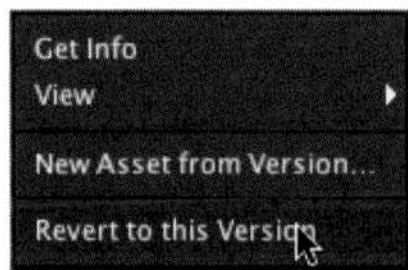

If you choose to revert, any later versions will be removed and that version will become the current version and be moved from the Versions device to replace the primary representation.

2 Close the info windows and quit the client application.

Updating a Checked-Out Asset

When you have an asset checked out, you can check it in but keep working on it. For this, there is another option called Update Asset, which will update the asset in Final Cut Server and create a new version, but keep the asset checked out.

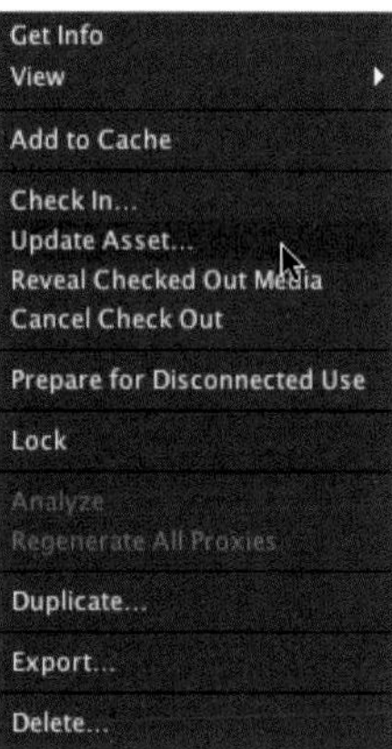

Lesson Review

1. What is an edit-in-place device?
2. When are Edit Proxies created?
3. Is it possible to define where Final Cut Server caches media?
4. If a media asset is not on an edit-in-place device, what must a user do before it can be used in Final Cut Pro?
5. What happens when a user reverts to a previous version of an asset?

Answers

1. An edit-in-place device is one in which the server and clients have access to the media in the same path. Typically this would be an Xsan device, but it is also possible to configure other devices as edit-in-place.
2. If Edit Proxies are enabled, they will be created for assets contained in a Final Cut Pro project when the project is added or checked in.
3. Yes. The location of this cache is determined in the Preferences window on the client application. You can also set the maximum cache size.
4. The asset must first be cached on the local machine. It can then be dragged into Final Cut Pro.
5. The selected version replaces the primary representation, and all later versions are removed.

7

Goals

Work with Motion and Soundtrack Pro

Work with graphics applications

Map metadata

Work with Podcast Producer

Export XML

Lesson 7

Integrating with Applications

This lesson is dedicated to scenarios and subjects that fall outside of the normal deployment of Final Cut Server but can be utilized to extend its flexibility and functionality. Some of these will be performance enhancements for accelerating Final Cut project analysis and speeding up database access.

Working with Final Cut Studio Projects

When uploading Motion, DVD Studio Pro, and Soundtrack Pro audio projects, Final Cut Server may recognize the project files as bundles in order to manage them as a single asset. During the upload process, you will see a message reminding you that Final Cut Server will not automatically account for any media files that are not within that project file's bundle.

In an Xsan environment, you can use Final Cut Server to ease collaboration between the studio apps, leveraging the fact that the media used in the projects is stored on an edit-in-place device (the Xsan) and is accessible to each Xsan client.

ADMIN ▼

Preparing for Motion Projects

In this exercise you'll create a workflow to allow collaboration between an editor using Final Cut Pro and a graphic artist using Motion. Usually, these people would be on separate machines connected to an Xsan, but as in Lesson 6, you'll simulate the Xsan on one machine by using a local directory as an edit-in-place device.

1. Open Finder and navigate to the /FCSvr/Leverage EIP directory. Create a new folder at that level named *Motion*.

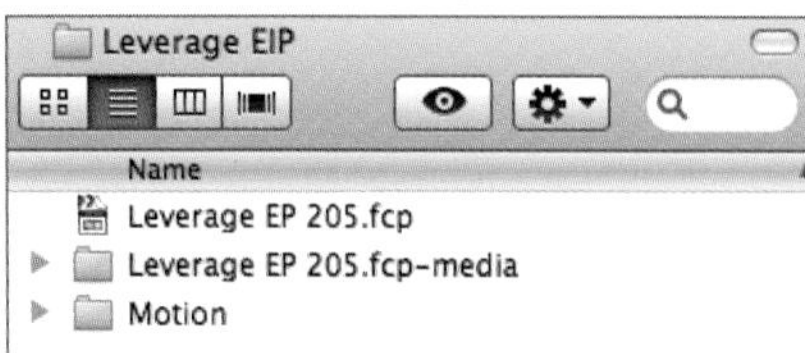

2 Go to the client application's Administration window. Choose Response from the pane on the left and click the Create button to make a new response. From the Response pop-up menu, choose Scan.

3 From the pane on the left, choose Create, and for Name and Description enter *Scan of Motion Projects*. Choose Scan from the left pane.

4 From the Scan Source pop-up menu, choose Leverage EIP (the device you created in Lesson 6). Click the Choose button and navigate to the Motion directory you created earlier, and then click Open.

5 From the Metadata Set pop-up menu, choose Project (since you're going to scan Motion projects). In the Asset metadata, choose "Ready for Editing" from the Status pop-up menu.

6 For Scan Mode, choose Add Only. For Entity Type, choose File. For Recursion Limit, enter *0* (for unlimited).

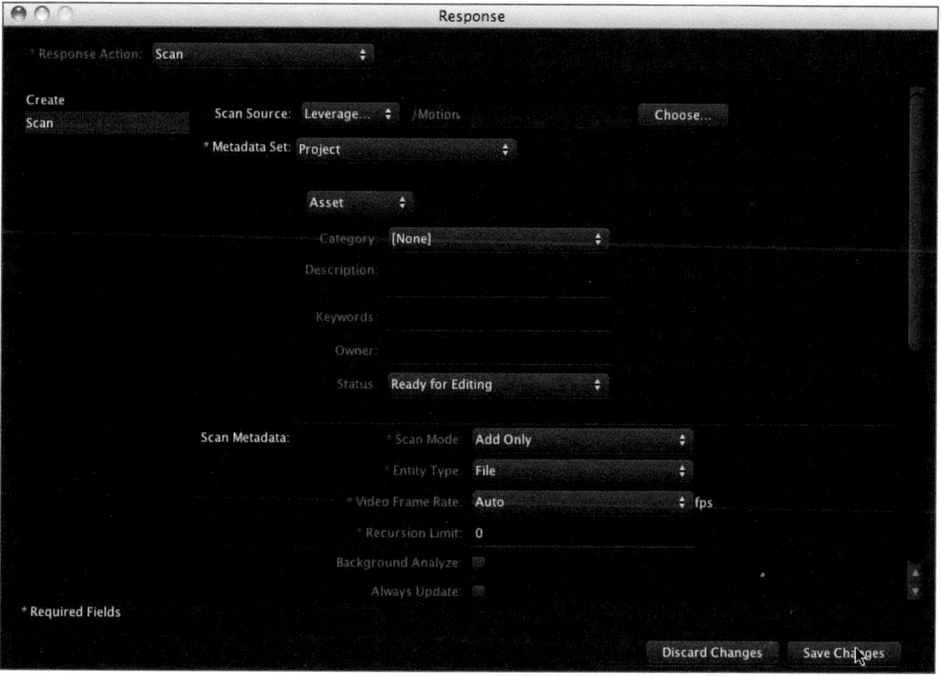

7 Add a Wildcard Include filter of **.motn*. Click Save Changes.

This ensures that only Motion projects are picked up by this scan.

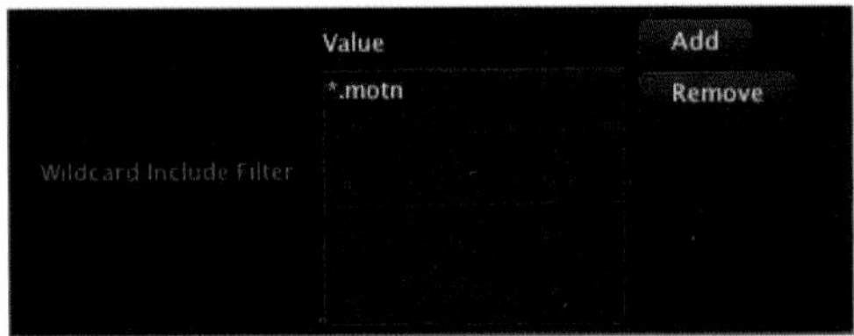

Now you need to add the scan to a schedule so it can be triggered.

8 Choose Schedule from the left pane and click the Create button. In the Schedule pop-up menu, choose Periodically. In the Name field, enter *Scan of Motion Projects*. Select Enabled. In the Response List, add the Scan of Motion Projects response. In the Scheduled Period (in minutes) field, enter *1*, which sets up the response to scan every minute. Click Save Changes.

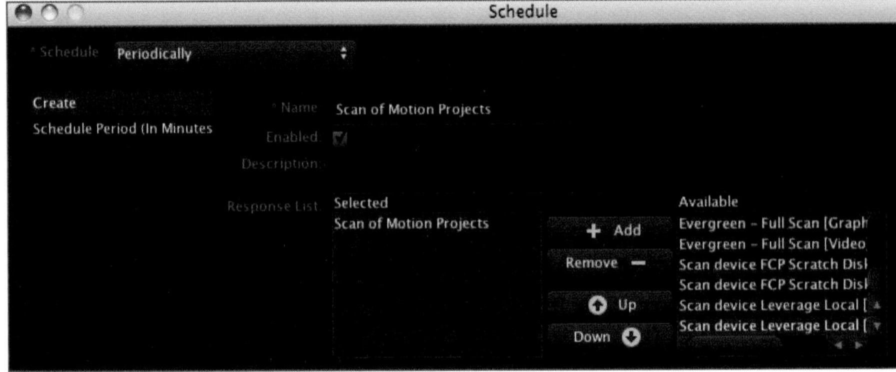

USER ▼

Roundtripping a Motion Project

You're now ready to open Final Cut Pro and send something to your Motion artist.

1 In Final Cut Server, find the Leverage Final Cut Pro project file you used in Lesson 6 and check it out to the CHECKOUT folder.

NOTE ▶ If you receive an "already exists" warning, click Yes. This is because you're overwriting the file you had previously checked out.

2 Open the project file in Final Cut Pro.

3 In the sequence, right-click the first instance of the clip **7-3** and choose Send To > Motion Project from the shortcut menu.

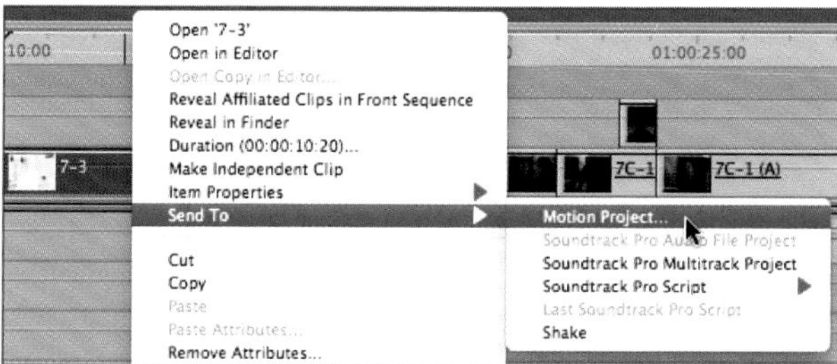

4 In the dialog that opens, in the Save As field, enter *Motion Project 1*. For the destination, select the Motion directory you created earlier. Deselect the Launch "Motion" checkbox. Click Save.

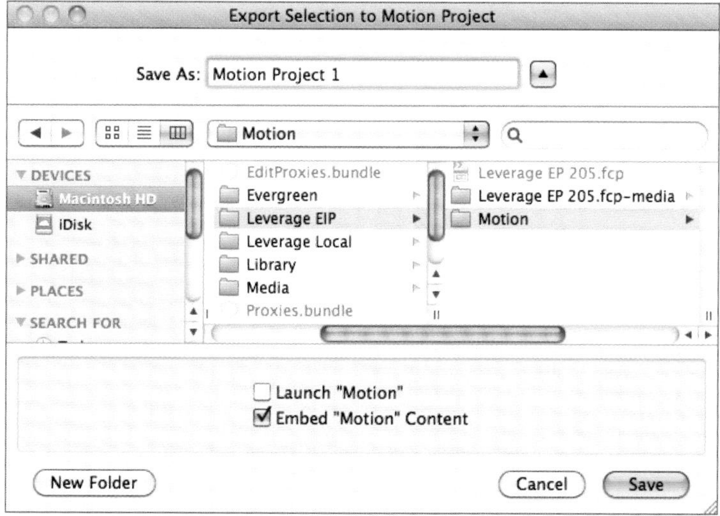

5 At some point in the next minute, Final Cut Server will scan Motion Project 1 and create an asset out of it. Search for *Motion* to find it.

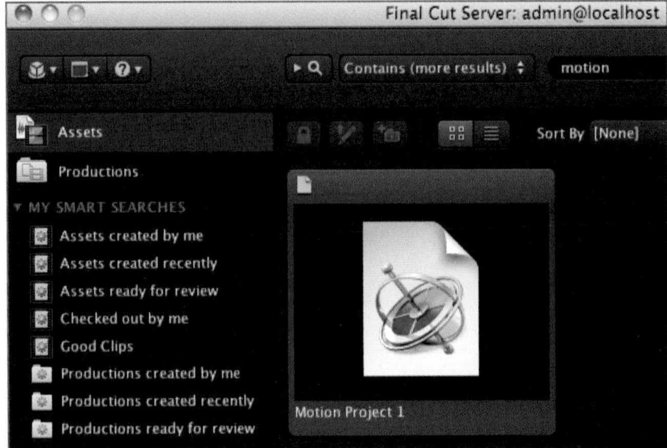

Now the graphic artist will check out the Motion project and make changes.

6 Right-click **Motion Project 1** and choose Check Out from the shortcut menu. Set the destination as the CHECKOUT folder on your desktop, and then click Check Out.

7 In the Finder, double-click the **Motion Project 1** file in the CHECKOUT folder.

8 Select the Text tool.

9 Enter the word *Leverage* somewhere in the canvas and press the Escape key.

10 Save the project, and quit Motion.

11 In the client application, right-click **Motion Project 1** and check it in. Enter some comments in the Version Comments field and then click Check In.

Final Cut Server has now copied **Motion Project 1** back to the original location.

12 Go back to Final Cut Pro and you'll see that the work you've done in Motion is now reflected in the sequence.

Graphics work can be an ongoing iterative process between an editor and an artist. For this exercise, you'll continue the process by sending to Motion one more time in order to see how you can use versioning in Final Cut Server.

13 Check out the Motion project again, and then open it via the Finder.

14 Select the text if it's not already enclosed in a bounding box and drag the text to a new position.

15 If the HUD is not visible, click the HUD button at the top right of the interface.

16 In the HUD, drag the Size slider to resize the text. Click the color swatch to open the color picker. Choose a different color for the text.

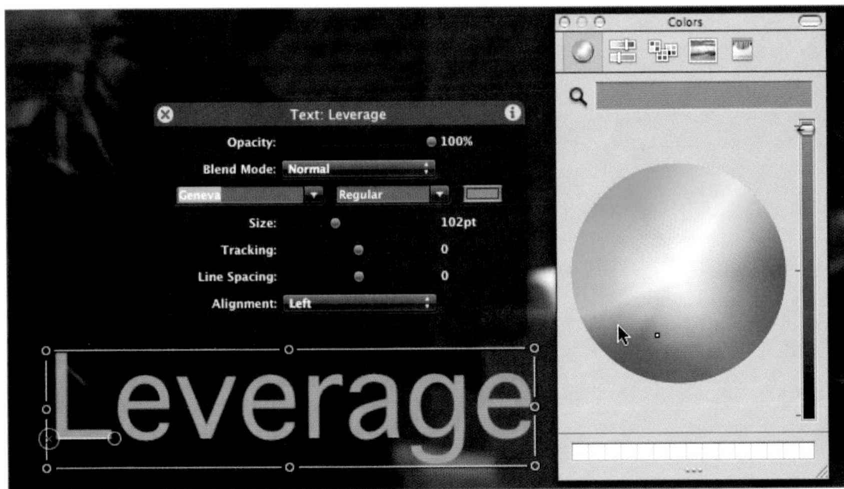

17 Save the project and quit Motion.

18 Check in the Motion project and enter some version comments.

19 Switch to Final Cut Pro to see the sequence update.

20 Once you've seen the change reflected in Final Cut Pro, in Final Cut Server double-click **Motion Project 1** and click the Versions tab.

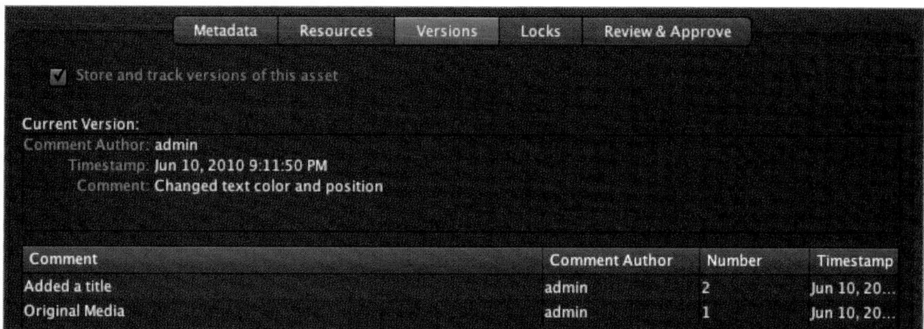

You'll see the original project, as well as the two iterations of changes that you've made in Motion.

21 Right-click version 2 and choose "Revert to this Version" from the shortcut menu.

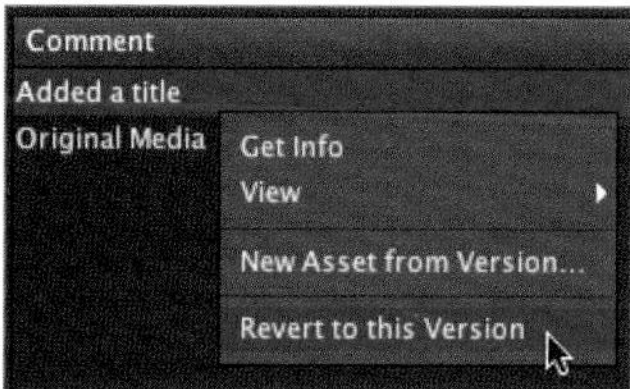

22 A dialog opens, stating that all later versions will be removed. Click Revert.

23 Go back to the sequence in Final Cut Pro. You'll see that the original Motion work is now visible.

24 Save the Final Cut Pro project, quit Final Cut Pro, and check in the Final Cut Pro project file.

This should give you a fundamental understanding of how Final Cut Server can assist in the workflow between editors and motion artists in an Xsan environment. Such a workflow can be extended by including additional responses that may do things such as send notification emails to the relevant staff.

Working with Graphics Applications

In some cases, you may want Final Cut Server to catalog the graphics or images that have been created on users' machines using programs such as Aperture or Adobe Photoshop. Each of these applications supports metadata in the file, and it's often very useful to have that metadata added to Final Cut Server automatically.

Three of the common standards to store metadata in graphic files are IPTC, EXIF, and XMP.

The *IPTC Information Interchange Model* (IMM) standard was adopted in the early 1990s to provide metadata fields in digital image files. IPTC fields are usually stored in the header of a file. The model has been extended to include photo metadata. It can be embedded in JPEG- or TIFF-formatted files.

The *Exchangeable Image File Format* (EXIF) is a standard used to describe metadata, most commonly for digital cameras.

The *Extensible Metadata Model* (XMP) is a newer standard that was created by Adobe Systems. It supports a wider variety of files and is extensible—it can accommodate existing metadata schemas and can easily be extended with new ones. It can be used in JPEG, JPEG 2000, GIF, PNG, TIFF, Adobe Illustrator, PSD, PostScript, and Encapsulated PostScript as well as PDF and HTML.

Final Cut Server is able to read each of these standards, and can also understand metadata in other file formats such as QuickTime. The metadata from these files can be seen in Final Cut Server assets through a process called "mapping."

ADMIN ▼

Modifying the Graphic Metadata Set

In the next few exercises, you'll use a watch folder to import graphics into Final Cut Server. At first you'll use built-in metadata and mappings, and then you'll see how to create new fields and mapping to match fields that are available in the source applications.

When installed, Final Cut Server sets up a few watchers in the Watchers device. One of those watchers is the Graphic watch folder. Before you use this watcher, you'll modify the Graphic metadata set to recognize an asset's EXIF metadata.

1 In the Administration window, click the Metadata Set pane.

2 Double-click the Graphic metadata set.

   This is the default metadata set applied to files dropped into the Graphic watcher.

3 Select the "Photo Info - EXIF" group from the Available list and add it to the set. Click Save Changes.When installed, Final Cut Server sets up a few watchers in a Watchers device. We'll use one of these to get graphics into the system.

   This group contains a number of fields that have mappings—such as Focal Length and Camera Model—to what you would see in a program such as Aperture.

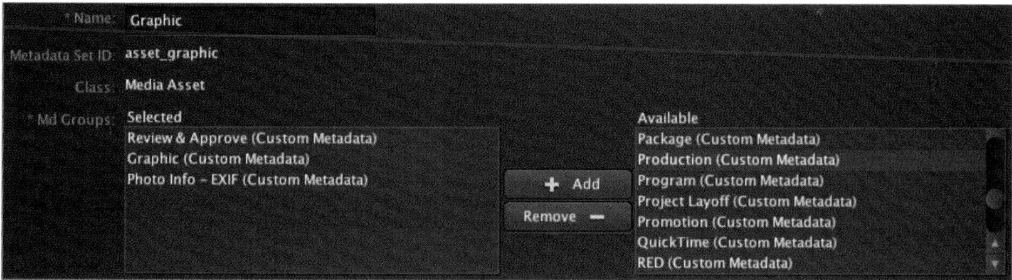

USER ▼

Cataloging Photos

Now that you've modified the metadata set, you're ready to add an image to the watch folder and see what happens.

1 Navigate to the FCS_Book_Files/Photos folder. Drag the **North Bondi Sunset.jpg** asset into the Macintosh HD/FCSvr/Watchers/Graphic folder.

   After a short while you should see some activity in the Jobs window.

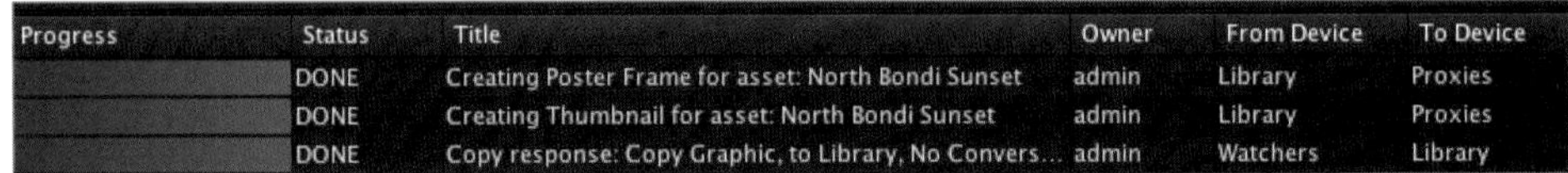

Progress
Status
Title
Owner
From Device
To Device
DONE
Creating Poster Frame for asset: North Bondi Sunset
admin
Library
Proxies
DONE
Creating Thumbnail for asset: North Bondi Sunset
admin
Library
Proxies
DONE
Copy response: Copy Graphic, to Library, No Convers...
admin
Watchers
Library

2 Search for the asset and double-click it. Go to the Photo Info – EXIF metadata group, where you'll see all of the metadata that has come through with the file.

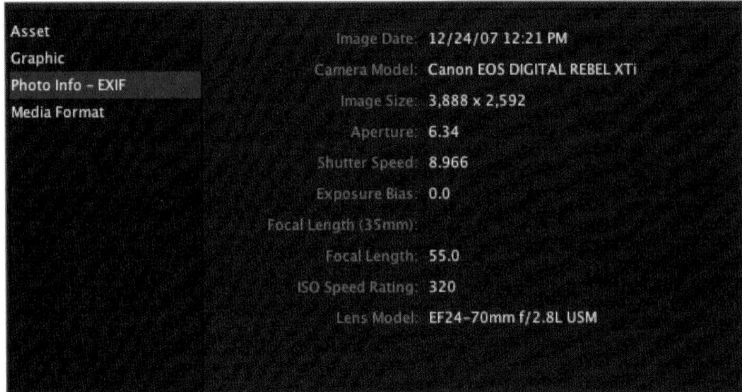

Metadata Mapping

You saw in the previous exercise how it's possible for metadata in a file to be mapped to a metadata field in an asset. You used some built-in mappings that ship with Final Cut Server. In this example, you'll see how to set up custom mapping.

Metadata can be embedded into files a number of ways. Digital cameras will write metadata to the files they create, and these can be modified in image editing applications.

The following image shows the metadata in Aperture. Certain fields are editable:

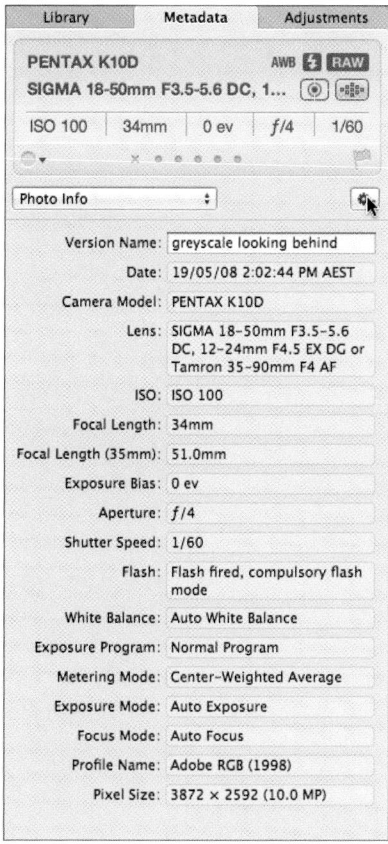

XMP and IPTC metadata can also be seen and edited in Adobe Photoshop CS3 or later. You can see this by choosing File > File Info.

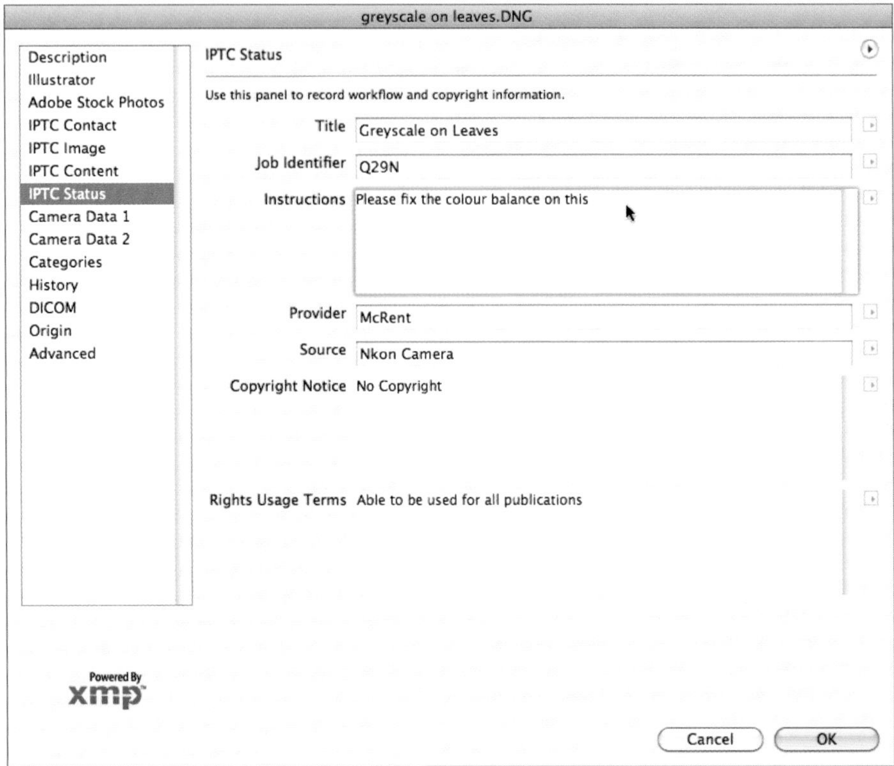

Most of the fields are already mapped to internal Final Cut Server fields. In order to see them appear in an asset, you'll need to add the corresponding field into a metadata group that belongs to the metadata set you want to use.

ADMIN ▼

Configuring Custom Metadata Maps

In this exercise you'll map the Instructions field into a new field—called My Metadata Field—that you'll create in Final Cut Server.

1 Go to the Administration window. Choose Metadata Field from the pane on the left, and click the Create button to make a new field. In the Name field, enter *My Metadata Field*.

2 From the Data Type pop-up menu, choose Unicode String. In the Description, enter *A metadata field*. From the Category pop-up menu, choose None. Click Save Changes.

You need to add the field to a metadata group so it can be seen in an asset. You'll add it to the Photo Info – EXIF group you used earlier.

3 In the Administration window, choose Metadata Group, and search for *photo info*. Open the Photo Info – EXIF group. In the Available List, add My Metadata Field, and then click Save Changes.

4 Quit and restart.

5 Choose Metadata Map from the pane on the left, and click the Create button to make a new map.

6 From the From Field pop-up menu, choose Instructions (XMP Metadata).

7 From the To Field pop-up menu, choose My Metadata Field (Customer Metadata).

8 In the Priority field, enter *3*.

NOTE ▶ Mappings with a lower number in the Priority field take priority when two or more fields are mapping to the same field.

9 Select the "Two way map" checkbox. Click Save Changes.

This option ensures that the mapping occurs in both directions when a file is analyzed by Final Cut Server and an asset is created, as well as when one is exported.

USER ▼

Viewing Custom Metadata

Now that you've added your field and set up the mapping, you can add a file to see it in action.

1 Go to the FCS_Book_Files/Photos folder. Drag **greyscale on leaves.psd** into the graphic watcher you used earlier in the lesson. After a short while, you should see some activity in the Jobs window.

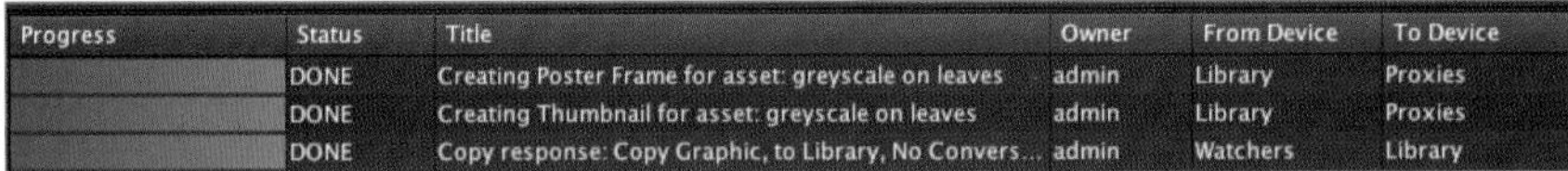

2 In the Asset window, search for *greyscale*. Double-click the asset to show the metadata, and select the Photo Info – EXIF Metadata group.

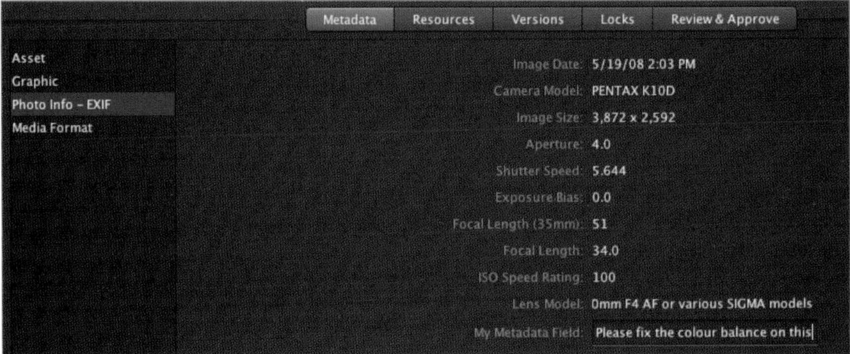

The metadata has now been mapped into your new field.

This is a simple example of how you can map metadata from external to internal fields. This can be extended in a production scenario as you build up your own custom metadata groups and watch folders or scans. Graphic artists can enter metadata quickly and easily for the file in the application they are using, as you have seen for Aperture and Photoshop. This will save them from having to re-enter metadata in Final Cut Server.

With two-way mapping you can take advantage of this the other way: A user could enter metadata for an asset in Final Cut Server that could then be exported to be used by a graphic artist who is able to view the metadata in his or her graphics application.

NOTE ▶ Two-way mapping does not alter the original file; the metadata is only embedded upon export.

Exporting Metadata as XML

In the previous examples, you've seen how metadata can be embedded into files and used by Final Cut Server. Sometimes it's desirable to have the metadata for an asset exported as a separate file, which might then be parsed by another application. Final Cut Server is able to do this by using the Write XML response.

ADMIN ▼

Preparing Final Cut Server

In this example you'll export an image and a corresponding XML file to an output location. The export will occur when the Status field is changed to Export.

1 In the Finder, create a new directory named *Output* in /FCSvr/Leverage EIP.

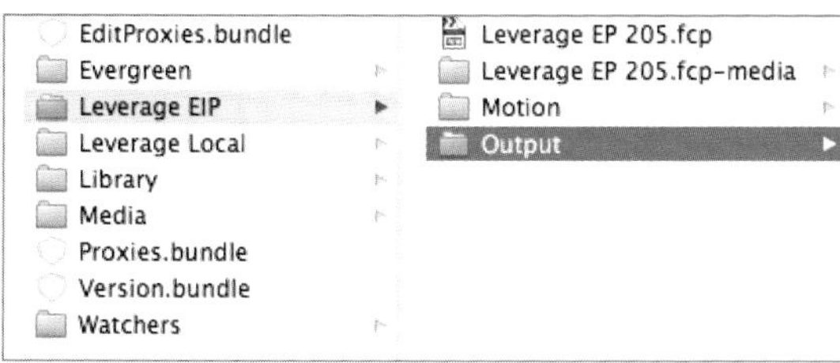

2 Go to the Administration window. Choose Response from the pane on the left, and click the Create button to make a new response.

3 Select Write XML as the Response Action. Set the title to *Write XML*. Set the Destination to Leverage EIP, and then select the Output directory. Select the "Use ID for filename" checkbox. Click Save Changes.

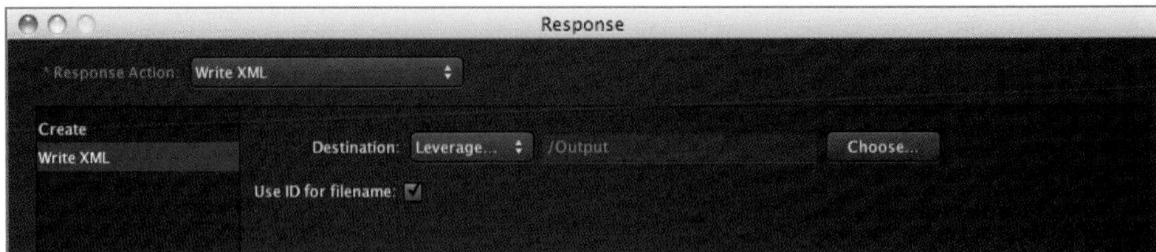

4 Create another response. This time, select Copy as the Response Action. Set the name to *Copy Image to Output*. Set the Destination to Leverage EIP, and then select the Output directory. Set the Transcode Setting to No Conversion. You do not need to select the Create Asset checkbox, as you are exporting a copy of the original image and do not need to keep a copy in the catalog. Click Save Changes.

You now need to add an Export entry to the Status metadata field lookup.

5 Choose Lookup from the pane on the left, and search for *Asset Status*. Double-click it to edit the lookup. Enter *Export* in the Name and Value fields, and click Add. Click Save Changes.

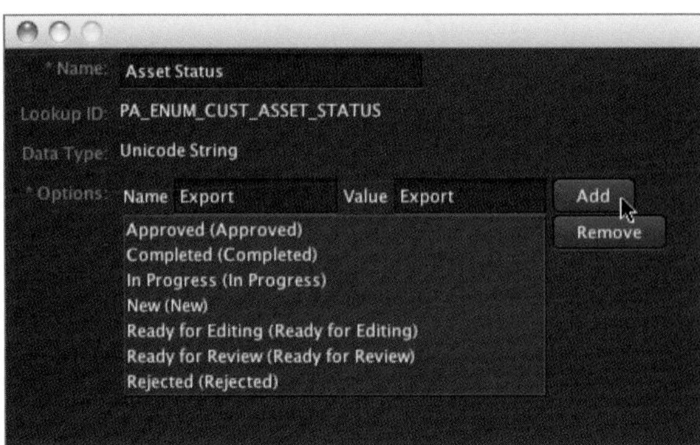

6 Restart the client application.

Entries in a lookup are cached when the client application starts, so to be able to see the new value in the Status field you will now need to quit the client application and start it again.

7 To trigger the two responses, you'll create a subscription. Go back into the Administration window and choose Subscription from the pane on the left, and then click Create to make a new subscription.

8 In the first window, in the "Subscribe to" pop-up menu, choose Asset. Enter *Export XML and Image* in the Name field. Select Enabled. Choose Modified from the Event Type Filter list. In the Response List, add the Write XML and "Copy Image to Output" responses.

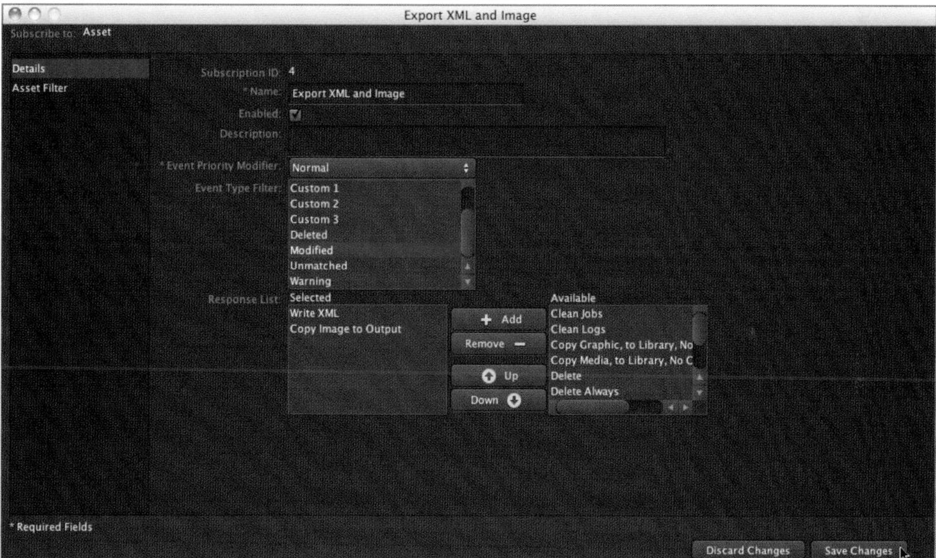

9 In the Asset Filter pane, choose Equals and Graphic for the Metadata Set. In the Status field, choose Equals and enter *Export*, and select the "Trigger if changed" checkbox to ensure that the response will trigger when this particular field has been modified. Click Save Changes.

USER ▼

Writing XML Based on Status

You can now test this subscription on a graphic asset. Exit the Administration window and go to the Asset window.

1 In the Asset window, search for *greyscale* and double-click the asset. Change the Status to Export. Click Save Changes.

If you look in the Jobs window, you will see that the Write XML response has been triggered.

2 Go to the Output directory you created, and open the XML file in TextEdit. It will look something like this:

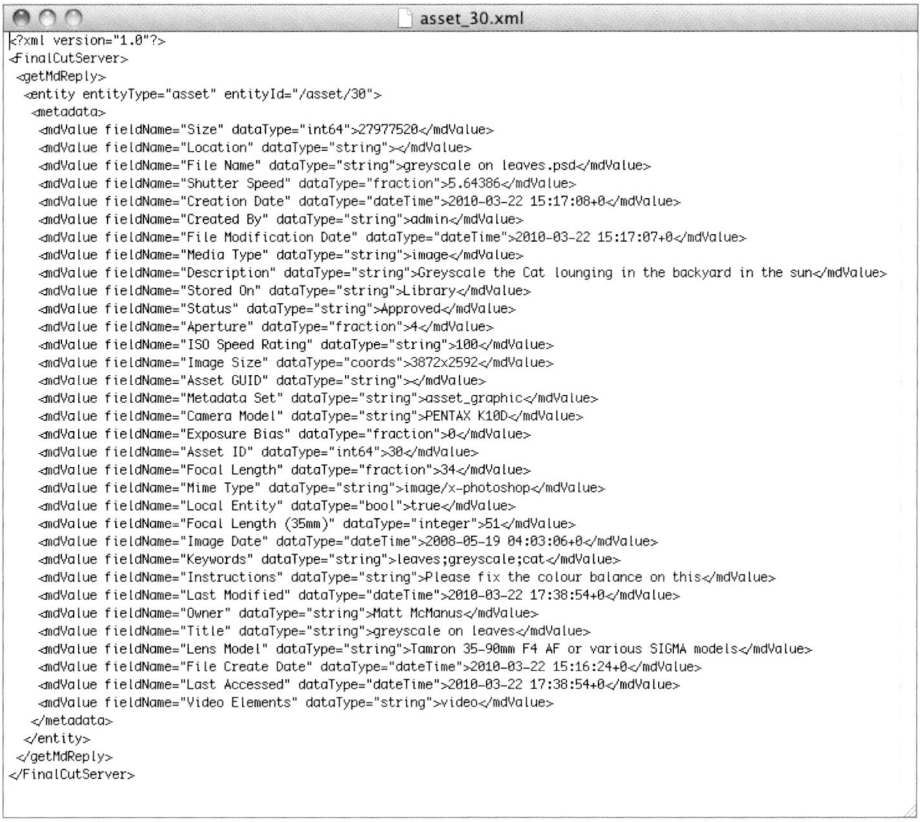

asset_30.xml


```
<?xml version="1.0"?>
<FinalCutServer>
 <getMdReply>
  <entity entityType="asset" entityId="/asset/30">
   <metadata>
    <mdValue fieldName="Size" dataType="int64">27977520</mdValue>
    <mdValue fieldName="Location" dataType="string"></mdValue>
    <mdValue fieldName="File Name" dataType="string">greyscale on leaves.psd</mdValue>
    <mdValue fieldName="Shutter Speed" dataType="fraction">5.64386</mdValue>
    <mdValue fieldName="Creation Date" dataType="dateTime">2010-03-22 15:17:08+0</mdValue>
    <mdValue fieldName="Created By" dataType="string">admin</mdValue>
    <mdValue fieldName="File Modification Date" dataType="dateTime">2010-03-22 15:17:07+0</mdValue>
    <mdValue fieldName="Media Type" dataType="string">image</mdValue>
    <mdValue fieldName="Description" dataType="string">Greyscale the Cat lounging in the backyard in the sun</mdValue>
    <mdValue fieldName="Stored On" dataType="string">Library</mdValue>
    <mdValue fieldName="Status" dataType="string">Approved</mdValue>
    <mdValue fieldName="Aperture" dataType="fraction">4</mdValue>
    <mdValue fieldName="ISO Speed Rating" dataType="string">100</mdValue>
    <mdValue fieldName="Image Size" dataType="coords">3872x2592</mdValue>
    <mdValue fieldName="Asset GUID" dataType="string"></mdValue>
    <mdValue fieldName="Metadata Set" dataType="string">asset_graphic</mdValue>
    <mdValue fieldName="Camera Model" dataType="string">PENTAX K10D</mdValue>
    <mdValue fieldName="Exposure Bias" dataType="fraction">0</mdValue>
    <mdValue fieldName="Asset ID" dataType="int64">30</mdValue>
    <mdValue fieldName="Focal Length" dataType="fraction">34</mdValue>
    <mdValue fieldName="Mime Type" dataType="string">image/x-photoshop</mdValue>
    <mdValue fieldName="Local Entity" dataType="bool">true</mdValue>
    <mdValue fieldName="Focal Length (35mm)" dataType="integer">51</mdValue>
    <mdValue fieldName="Image Date" dataType="dateTime">2008-05-19 04:03:06+0</mdValue>
    <mdValue fieldName="Keywords" dataType="string">leaves;greyscale;cat</mdValue>
    <mdValue fieldName="Instructions" dataType="string">Please fix the colour balance on this</mdValue>
    <mdValue fieldName="Last Modified" dataType="dateTime">2010-03-22 17:38:54+0</mdValue>
    <mdValue fieldName="Owner" dataType="string">Matt McManus</mdValue>
    <mdValue fieldName="Title" dataType="string">greyscale on leaves</mdValue>
    <mdValue fieldName="Lens Model" dataType="string">Tamron 35-90mm F4 AF or various SIGMA models</mdValue>
    <mdValue fieldName="File Create Date" dataType="dateTime">2010-03-22 15:16:24+0</mdValue>
    <mdValue fieldName="Last Accessed" dataType="dateTime">2010-03-22 17:38:54+0</mdValue>
    <mdValue fieldName="Video Elements" dataType="string">video</mdValue>
   </metadata>
  </entity>
 </getMdReply>
</FinalCutServer>
```


You can see from this example that it's possible to export metadata as a separate file. To extend this further in a workflow, you might also add additional responses to the subscription—for example, a copy response to export a file as well as a "run external script or command" response in order to run a script that might parse the XML and do something useful.

Working with Podcast Producer

Snow Leopard Server included a new version of Podcast Producer that has the option to use Final Cut Server as a publish destination.

When creating a Podcast Producer workflow, you'll see an option to add Final Cut Server.

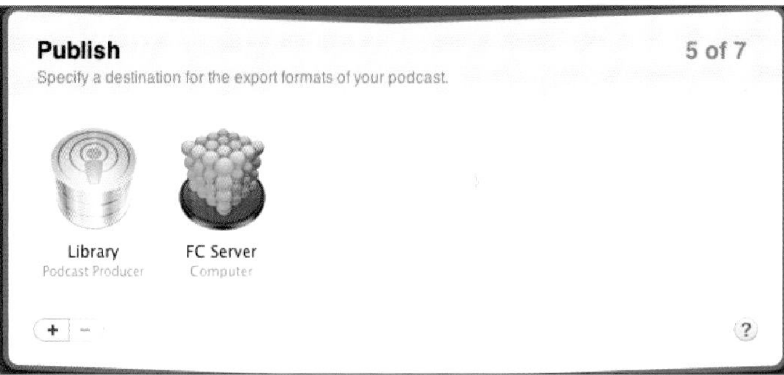

You can select the version of the podcast you want to copy, and then select a path to a watch folder in Final Cut Server. You'll need to configure this watch folder.

It's also possible to send files to Podcast Producer. It's beyond the scope of this book, but one possible way to do this is to use the "Run an external script or command" response in combination with the Podcast Producer command-line tools. See the manuals of each product for further information.

Lesson Review

1. What do you need to check when you upload a Motion, Soundtrack, or DVD Studio Pro project to Final Cut Server?
2. What are four image file formats that support metadata?
3. Where is the Final Cut Server metadata mapping functionality located?
4. What happens when two external fields are mapped to the same internal field in Final Cut Server?
5. What mechanism is used to export metadata from Final Cut Server in a separate file?

Answers

1. You need to check that any media linked to the projects is in a place accessible to other users.
2. JPEG, TIFF, PSD, and Digital RAW files support metadata.
3. In the Metadata Map pane in the Administration window.
4. The field with the lower priority number will take precedence.
5. The Write XML response may be used to export metadata as a separate file.

Goals

- Review and approve assets
- Use the Annotations window
- Configure metadata subscriptions
- Configure email notifications
- Output annotations as XML

Lesson 8

Reviewing and Approving

The purpose of a Review and Approve workflow is streamlining the process of reviewing content, adding editorial comments, and then approving or rejecting the final content. Final Cut Server has a built-in template for a Review and Approve workflow that uses metadata fields, notifications, and subscriptions. Subscriptions are Final Cut Server processes that watch for changes in metadata, and then trigger a response. The template is a jumping-off point for molding Final Cut Server around your existing workflow. After you've used the application continuously for awhile, you'll find that you will come up with new ways to modify Final Cut Server to decrease the number of steps and the amount of time taken for your process.

In a common editorial workflow, a producer makes comments and choices about a piece of raw/ingested media, and then the editor makes cuts based on the comments. In the past, this was done over the phone, on notepads, and in other less-than-optimal fashion. With Final Cut Server, this entire process can be standardized, internalized in one application, and simplified so that the editor and producer can spend more time on their content than on the process.

The built-in template is easily customizable. In this lesson, you'll walk through the built-in template, customizing it to suit your project; and then you'll step through an actual workflow.

ADMIN ▼

Adding New Users for Review and Approve

Throughout the course of this lesson, you'll work with accounts for four users: an administrator, an editor, a producer, and an executive producer. If you're in a midsize to large environment that has directory services deployed, adding users to those services is outside the scope of this book. You can reference the Mac OS X Server Directory Service manual for step-by-step instructions on adding users to Open Directory. If you need to use Active Directory users, please consult the *Final Cut Server Administrator Guide* for step-by-step instructions on how to configure your system for integration with Active Directory.

You'll learn how to add local users using the Accounts pane in System Preferences. These accounts will be used during the Review and Approve process with email notifications, and also in using annotations. These are just example users and should not be used in production unless they're properly secured.

1 Open System Preferences, click Accounts, and then click the lock on the bottom to authenticate yourself as an administrator.

2 Enable Fast User Switching by clicking Login Options at the bottom left of the Accounts pane and selecting the checkbox for "Show fast user switching."

3 Click the Add (+) button to add the producer. For New Account, choose Standard (you don't want this user to administer the server). For Full Name, enter *Producer*; for Account, enter *producer*; for Password, enter *producer*; and then re-enter the same password for Verify. When you're finished, click Create Account.

4 Click the Add (+) button to add the Executive Producer user. For New Account, choose Standard; for Full Name, enter *Executive Producer*; for Account, enter *execproducer*; for Password, enter *execproducer*; and then re-enter the same password for Verify. When you're finished, click Create Account.

5 Click the Add (+) button to add the Editor user. For New Account, choose Standard; for Full Name, enter *Editor*; for Account, enter *editor*; for Password, enter *editor*; and then re-enter the same password for Verify. When you're finished, click Create Account.

6 Click the Add (+) button to add the Producer group. For New Account, choose Group; for Full Name, enter *Producer,* and click Create Group. Select Producer

from the Group pane on the left, and select the checkbox next to Producer in the Membership pane on the right.

7 Click the Add (+) button to add the ExecProducer group. For New Account, choose Group; for Full Name, enter *ExecProducer*, and click Create Group. Select ExecProducer from the Group pane on the left, and select the checkbox next to Executive Producer in the Membership pane on the right.

8 Click the Add (+) button to add the Editors group. For New Account, choose Group; for Full Name, enter *Editors*, and click Create Group. Select Editors from the Group pane on the left, and select the checkbox next to Editor in the Membership pane on the right.

Your system is now configured with the correct users to be able to demonstrate a successful Review and Approve workflow.

ADMIN ▼

Investigating the Built-In Review and Approve Template

Final Cut Server ships with a default metadata template that supports Review and Approve right out of the box. All the relevant fields for Review and Approve are stored in the Review and Approve metadata group. It's built into all of the default asset metadata sets and has some default fields that will work for most shops. If you're making your own custom metadata sets and want to use the Review and Approve functionality, it's important to add the Review and Approve metadata group to your custom metadata set.

Before you dive into customizing Review and Approve, let's take a look at Final Cut Server's built-in functionality and the automations that come installed by default. The default automation is a metadata subscription that sends out an email based on a change in the Status pop-up menu.

1 To begin, open the client application and log in as an administrator.

2 Search for the asset **SR0191 010 Jackpot** and double-click it to bring up the asset info window.

3 Click the Review & Approve button at the top of the asset info window to show the built-in Review and Approve functionality.

By default, Final Cut Server ships with a Status field that's tied to a status lookup that has some general steps in a workflow. It has a field called "Editor's comments," where editors and others can leave comments regarding the asset, and a field called "Required reviewers," where you can enter email addresses to be automatically notified when someone changes the Status field (this is tied to the default subscription we talked about earlier). We'll look at this function in detail in the next exercise.

4 Click the Status pop-up menu, and note the workflow steps that are available. Currently, only the Ready for Review field is tied to a subscription (which sends out an email with the editor's comments).

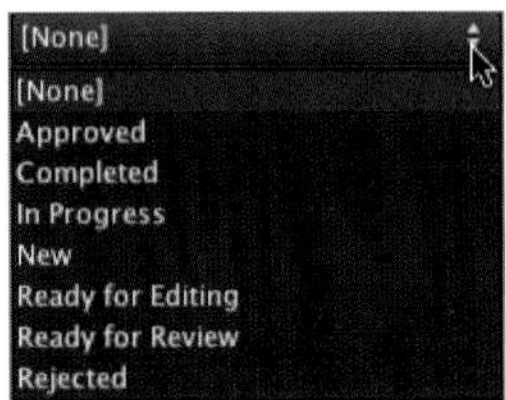

5 Close the asset info window, and quit the client application.

You're now going to investigate the built-in automation that sends out emails of the editor's comments. The simplest way to investigate this is through System Preferences.

6 Open System Preferences, go to the Final Cut Server pane, and authenticate.

7 Click the Automations button at the top of the pane, and select the Assets Ready for Review [Email] response.

8 Enable the response and then double-click to edit.

In the Metadata area, note that there are two steps in the metadata subscription. Final Cut Server treats this in a hierarchical manner, in this case with the Status metadata field being changed (instead of always match, which is usually not a good idea as it can lead to infinite loops). After the first step in the subscription is satisfied, the second step looks for the Status metadata field to match Ready for Review. If both these criteria are met, the subscription will trigger the responses associated with it (which you'll see in the next dialog). If you wish to add more criteria for the subscription, click the Add (+) button.

In short, this metadata subscription will be satisfied if a user chooses Ready for Review from the Status pop-up menu in the Review & Approve pane of any asset inside Final Cut Server and then clicks Save.

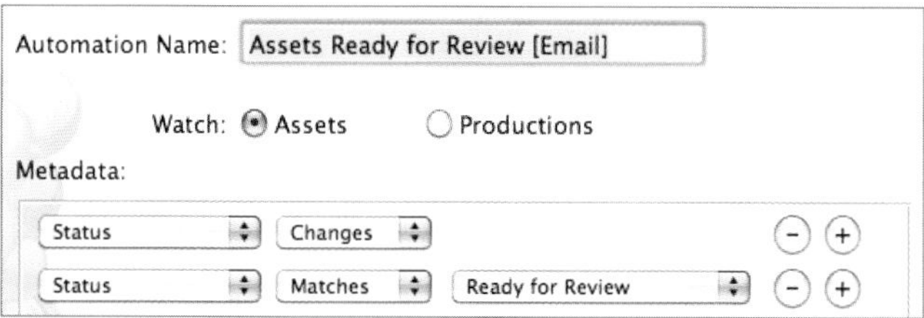

9 After reviewing the information, click Continue to go the Response window.

The Response window opens to assist you in adding responses to your subscription. These responses represent what occurs when the criteria in your metadata subscription are met.

In the list on the left, you can see that you've added one response: Email response. The Email response settings are on the right of the window and display the configuration of the actual email that will be sent when a user triggers the subscription (by changing the Status metadata field to Ready for Review).

One of the most powerful features of subscriptions is the ability to add metadata from the assets that triggered the subscription directly into the response.

The Review & Approve pane of an asset has a "Required reviewers" metadata field. In the Email pane of the Response window, the To field has "[Required reviewers]" inside it, and the Subject has "[Title]" in it. You can insert any metadata field information into a response by putting brackets around the EXACT field name, or by choosing a field from the Insert Metadata Field pop-up menu at the bottom of the Email pane.

Before clicking Continue, let's add a dummy account to the From field. Some organizations maintain tight control over email, and this account might have to actually exist, while in smaller shops it's commonplace to make a fake email address.

10 Type *fcsvr@pretendco.com* in the From field.

Next, let's add the "Editor's comments" field from the Review & Approve pane so that you can add those comments automatically to the email.

11 In the body of the email response, press Return to add a space below the first line, and then choose "Editor's comments" from the Insert Metadata Field pop-up menu.

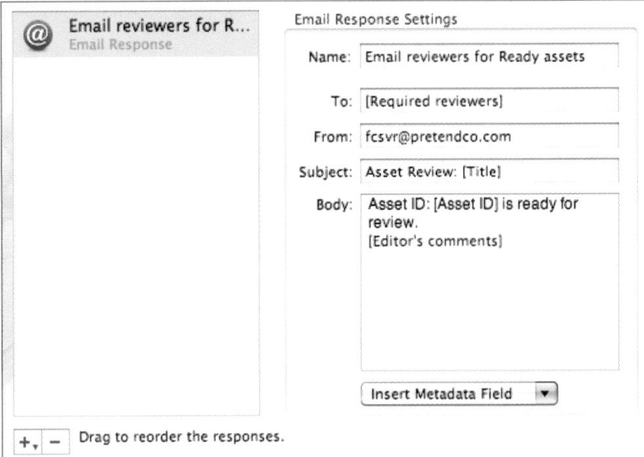

12 When you're finished, click Continue to go to the Summary dialog, and then click Done to save your update automation.

USER ▼

Triggering the Review and Approve Functions

Now that you have investigated how the Review and Approve default template and automations work, let's enable them and trigger their functionality. You'll add a reviewer, changing the Status metadata field to Ready for Review, and trigger the email response that is set up by default.

1 Open the client application and log in as the administrator.

NOTE ▸ The reason you're logging in as an administrator here is to view the Final Cut Server logs so that you can see your email response happening. If you do not have an SMTP server available, the log will provide proof the automation works.

2 If you didn't configure an SMTP (email) server during the initial installation of the server application and one is available, open the Administration window and enter the SMTP server address provided by your network administrator.

3 In the asset pane of the client application, do a search for *SR0191 010*. Double-click the resulting asset to open the asset info window, and click the Review & Approve button at the top.

Now you'll enter some information to trigger your subscription and have your email sent.

4 From the Status pop-up menu, choose Ready for Review. For "Editor's comments," add some information, such as *This clip is terrific! What do you think?* In the "Required reviewers" field, enter the email address of the producer. For this example, let's use *producer@pretendco.com*. When you're finished, click Save Changes, which will trigger your metadata subscription and send out an email with the information from the asset.

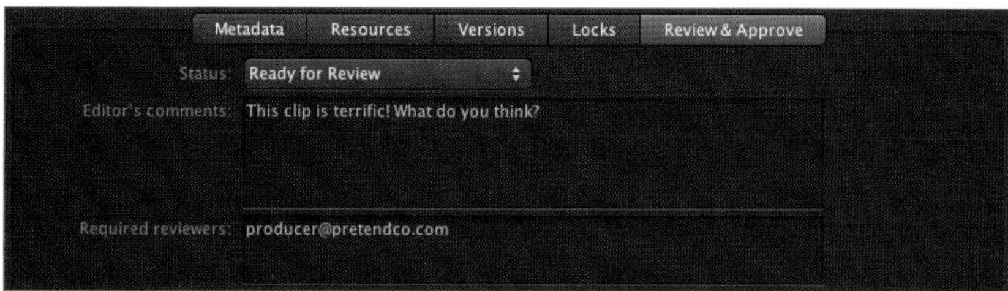

If SMTP is available and configured correctly, the email is sent, and you will be able to open Mail on your client computer and view the email that was sent from Final Cut Server.

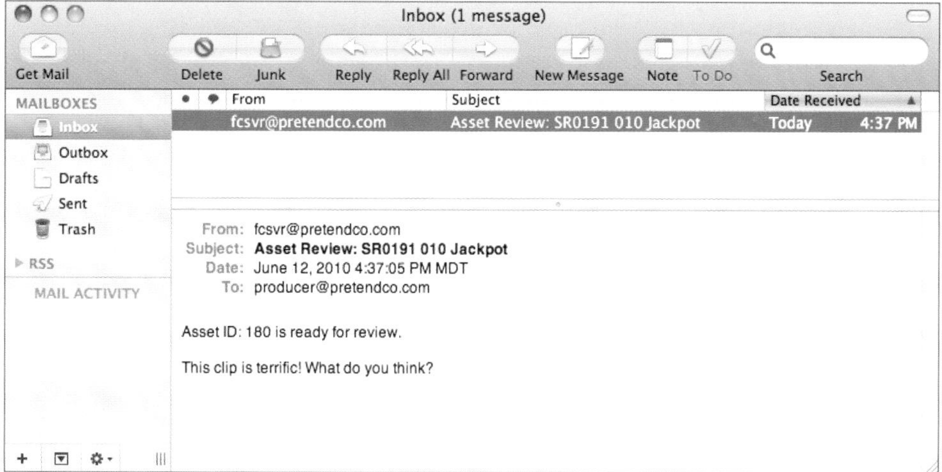

5 You can also check the status of an email response through the Log pane of the Administration window. Choose Administration from the server pop-up menu. Choose Log from the list on the left. Enter *Email* for the search term, and click Search. Note that the topmost entries are your email response being sent and completed.

TIP If an STMP is not available or configured correctly, you will find a response failed error message in the log.

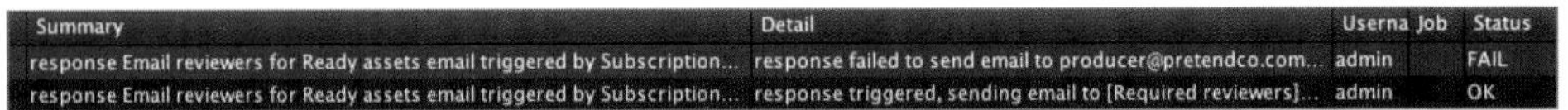

Summary	Detail	Userna	Job	Status
response Email reviewers for Ready assets email triggered by Subscription...	response failed to send email to producer@pretendco.com...	admin		FAIL
response Email reviewers for Ready assets email triggered by Subscription...	response triggered, sending email to [Required reviewers]...	admin		OK

6 Double-click the topmost entry to see the details that were logged as the email response was triggered by the change in metadata. The entry tells you the ID of the log entry, the time that the response was triggered, the name of the response triggered, and some detail about the actual response.

7 Close the Log entry window, but leave the Administration window open.

In the next part of the lesson, you'll learn how to customize the default template to better suit individual workflow needs.

ADMIN ▼

Customizing the Default Review and Approve Template

In the previous part of the lesson, you learned about the automations and metadata structure of the built-in Review and Approve functionality. For most shops, this default template will be a good starting point, but more customization might be needed. Many workflows have different steps and processes associated with them, so one of the most common modifications here is to change the values of the Status lookup. Also, some shops don't like allowing user entry for the "Required reviewers" field, as people have been known to mistype email addresses. This can be solved by using a lookup with this field. The lookup can list individual or group email addresses. This ensures that emails go out to the appropriate recipients.

More than likely, you will also need multiple text-entry fields for comments from each of the individuals in your workflow. Finally, one notification is usually not enough for a group of users. For the workflow to be successful, many people have to be notified at different points in the collaboration process. Also, you might not want certain individuals to have access to all the source material. For example, it's likely that you won't want your executive producer to see all the raw content in the catalog—just the finished assets that are clean and ready to be seen.

First, you will change the default values of the lookup to include two new steps in your workflow: Ready for Producer Review and Ready for Executive Producer Review. You'll then configure another lookup that has a list of all the email addresses for the required reviewers so that user error is less likely than it would be if you had to type in the addresses manually. You'll create two new text-entry fields for the executive producer and the producer to leave their comments. You'll set up two new subscriptions and email responses to notify the producer and executive producer when the status has been changed. Finally, you'll use and modify the default permission sets to restrict what the Producer and Executive Producer can see based on the Producer group.

1 In the Administration window, select Lookup on the left. Double-click the Asset Status to open the Asset Status window.

2 Highlight "Ready for Review" in the list of Lookup values and click Remove. This will remove the Ready for Review item from the Status pop-up menu in the Review & Approve pane.

3 Now you'll add two entries to the lineup. Enter *Ready for Producer Review* into the Name text-entry box and *Ready for Producer Review* in the Value text-entry box, and click Add. Repeat using *Ready for Executive Producer Review*. Adding these entries will make them available in the Status lookup and allow us to assign metadata subscriptions to them in order to send individual emails. Click Save Changes when you're finished to save the modified lookup.

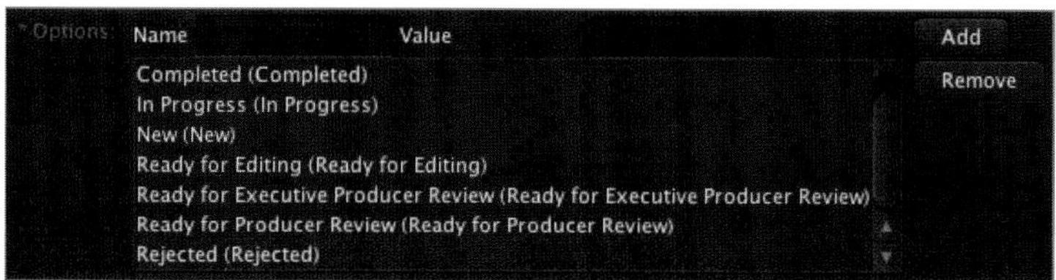

NOTE ▸ When it comes to lookups, the Name text is displayed in the UI, and the Value text is stored in the database. Ninety-five percent of the time, you want the name and value to match. On occasion, the Name and Value fields should not match. One example is an email address; the Name field would display the person's full name (Drew Tucker), but the Value would contain the actual email address (drew@somewhere.com).

Now that you've modified the existing status lookup, you can create one that contains the list of your reviewers and have the values be their email addresses.

4 In the Lookup pane of the Administration window, click the Create button to create a new lookup.

The Lookup window opens. You'll now add the email addresses of the required reviewers.

5 For Name, enter *Required Reviewers*. For Data Type, leave it at Unicode String (that allows us to use alphanumeric characters).

6 For Name, enter *Producer*; and for Value, enter *producer@pretendco.com* (or substitute the domain name given to you by your network administrator), and click Add.

7 Repeat this process using the following: for Name, *Executive Producer*; and for Value, *execproducer@pretendco.com* (again substituting your domain name). Click Save Changes when you're finished.

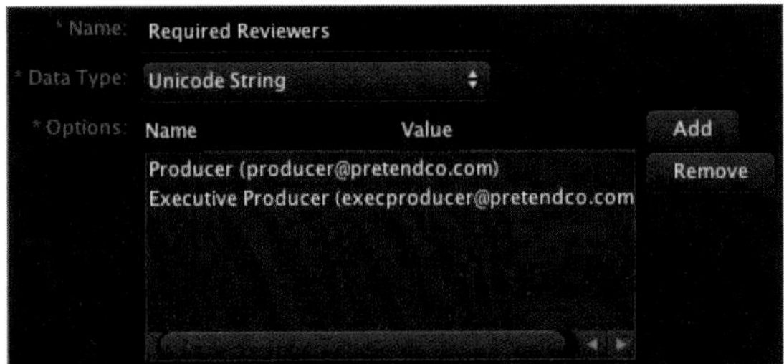

Associating Metadata Fields

Now that you've configured your new lookup values, you'll create two text-entry metadata fields and associate them with the Review and Approve metadata set. You'll then repurpose the existing "Required reviewers" metadata field to be associated with your new Required Reviewers lookup.

1 In the Administration window, choose Metadata Field from the list on the left. Click the Create button to make a new metadata field.

2 For Name, enter *Producers Comments*; for Data Type, leave Unicode String; for Description, enter *Comments made by the Producer*; for Category, choose None. In the "Display hints" area, select the Multiline checkbox to make the text-entry field allow multiline entries. You can leave the rest of the choices untouched; these display hints affect how

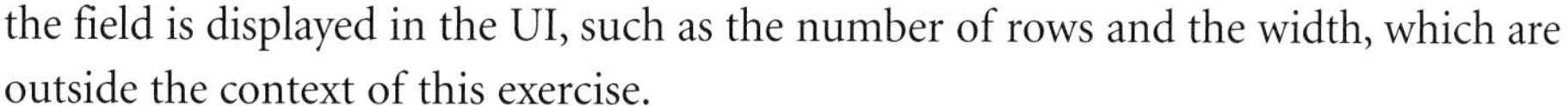

the field is displayed in the UI, such as the number of rows and the width, which are outside the context of this exercise.

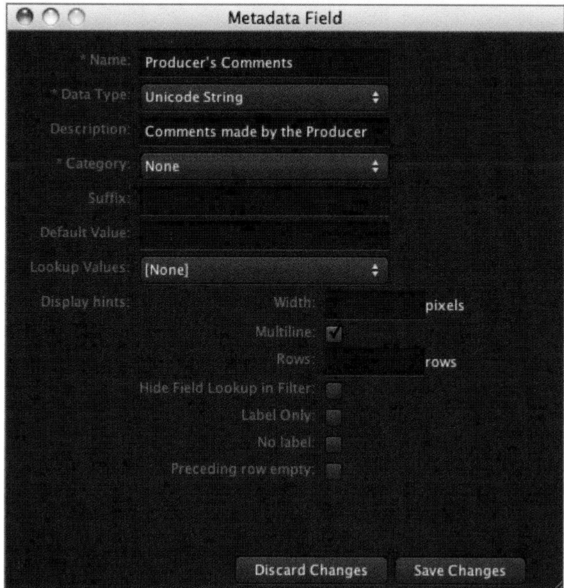

NOTE ▶ When you create all new metadata fields, the Category should *always* be set to None. The other values are internal to the application and will cause compatibility problems down the road.

3 Click Save Changes to return to the Metadata Field pane in the Administration window. Click Create again, and in the Name field, enter *Executive Producers Comments*; for Data Type, choose Unicode String; for Description, enter *Executive Producer Comments*; and for Category, choose None. Select the Multiline checkbox in the "Display hints" area. Click Save Changes when you're finished.

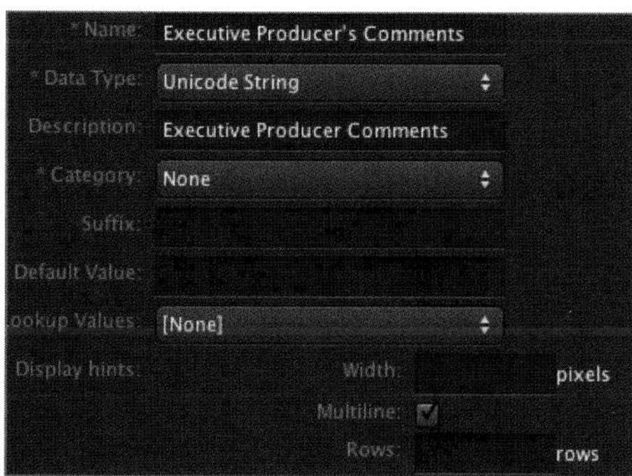

4 In the Administration window, choose Metadata Group from the list on the left, enter *review* in the search field, and click Search. Double-click the Review & Approve metadata group whose metadata group ID is PA_GRP_ASSET_APPROVAL.

NOTE ▶ In Final Cut Server, you will often have database entities with the same names, so it's important to look for entities that exactly match what you intend to change.

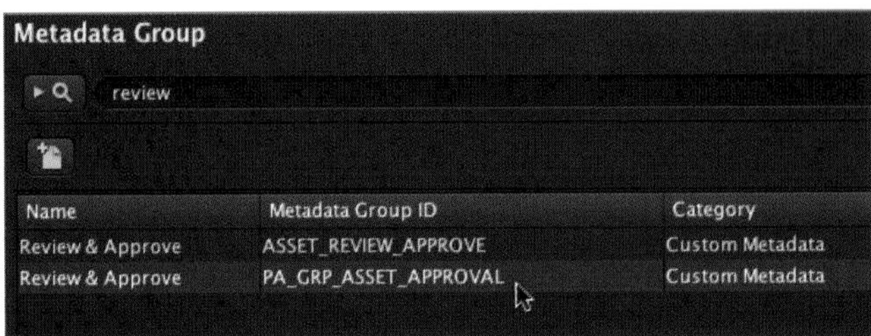

The Review & Approve (Custom Metadata) window opens. You'll now add the fields you configured in the previous steps in the Review & Approve metadata template.

5 In the list of Available fields on the right, click the Producer's Comments (Custom Metadata) field and click Add to move it to the list of Selected fields on the left. With the field highlighted on the left, click the Up button to move it directly under the "Editor's comments (Custom Metadata)" item.

6 Choose Executive Producer's Comments (Custom Metadata) from the list of Available fields on the right, and click Add to move it the list of Selected fields on the left.

7 With Executive Producer's Comments (Custom Metadata) highlighted in the Selected fields list, click Up to move it directly under the Producer's Comments (Custom Metadata) field. Both of the fields you created earlier are now available for metadata entry in the Review & Approve pane in each asset.

8 Click Required Reviewers. The Field Properties area appears below the Selected and Available lists. From the Field Lookup pop-up menu, choose Required Reviewers, the lookup you made earlier in the lesson. Also, under Field Properties, remove the entries for Width and Rows and leave them blank, and deselect Multiline. Associating the Required Reviewers lookup with the "Required reviewers" metadata field will make your pop-up menu available on the Review & Approve tab.

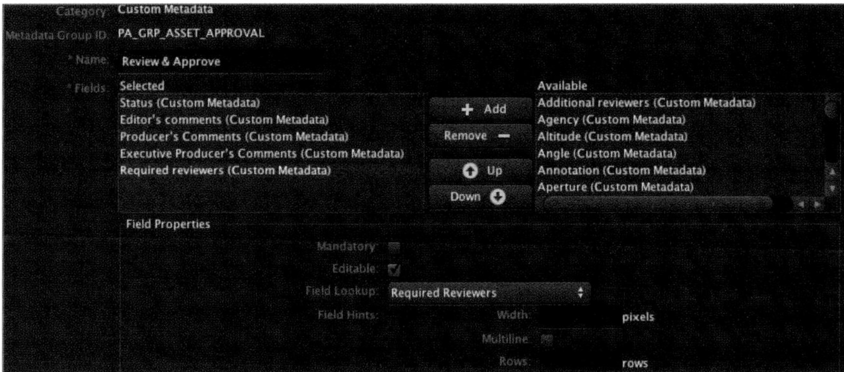

9 Click Save Changes when you're finished, and minimize the client application.

In the next exercise, you'll go to System Preferences to add the new subscription and email responses based on the new metadata structures we just created.

ADMIN ▼

Setting Up Email Responses for Status Change Notification

Now that you've configured the proper metadata template for your Review and Approve process, it's time to set up new email responses to notify your producers when the status is changed. You'll do this by creating two metadata subscriptions tied to the lookup values you created earlier in the lesson.

1. Open System Preferences, click the Final Cut Server icon, and authenticate.

2. Click the Automations button at the top, and then click the Add (+) button.

3. In the Automation Setup Assistant, select Metadata Subscription and click Continue.

4. Now you'll configure the metadata triggers you will subscribe to, in order to trigger your email response (which you'll create next). For Automation Name, enter *Email Response for Ready for Producer Review*; for Watch, select the Assets button. You're looking for a change in asset metadata, not production metadata.

5. In the Metadata area, choose Status from the first pop-up menu, and then choose Changes from the second pop-up menu. You choose Changes because you want to

trigger this email notification only if the Status is changed. Click the Add (+) button at the right of the Metadata area to add additional criteria.

6 In the new row that appears, from the first pop-up menu, choose Status; from the second pop-up menu, choose Matches; and from the third pop-up menu, choose Ready for Producer Review. Click Continue.

Now your metadata subscription has been configured to look for a change in the Status field to Ready for Producer Review. You'll configure the email response in the next dialog.

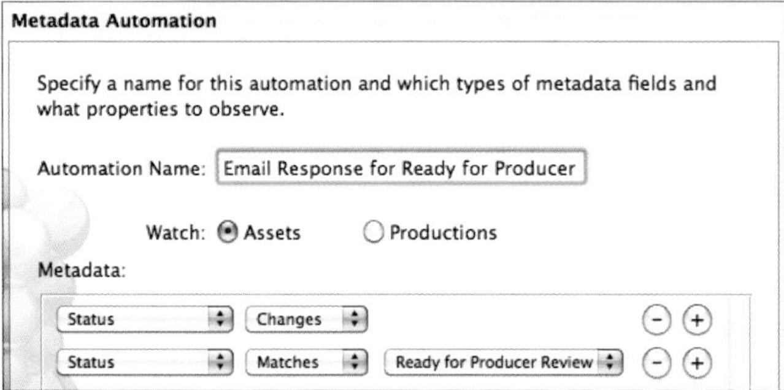

7 Click the Add (+) button at the bottom of the Responses dialog, and choose Email Response from the pop-up menu. For Name, enter *Email Response for Ready for Producer Review.*

8 Complete the remaining fields as shown in the following image. Use the Insert Metadata Field pop-up menu for the bracketed ([ ]) text items.

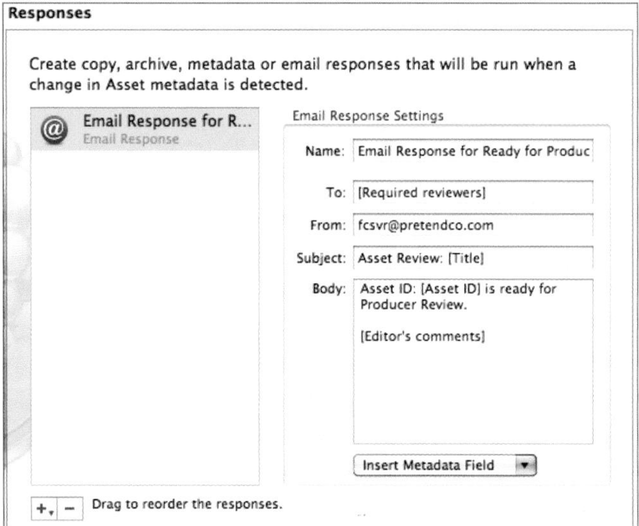

Your email response is now configured to send an email to the producer when "Ready for Producer Review" is chosen from the Status pop-up menu.

9 Click Continue after you have reviewed what you entered against the screen shot. Click Done on the next screen to finish adding your metadata subscription.

10 In the Automations pane, click the Add (+) button. In the Automation Setup Assistant, select Metadata Subscription and click Continue.

11 You will now configure the metadata triggers you will subscribe to, in order to trigger your email response for executive producer review. For Automation Name, enter *Ready for Executive Producer Review Email*; for Watch, select Assets. You're looking for a change in asset metadata, not production metadata.

12 In the Metadata area, choose Status from the first pop-up menu, and then choose Changes from the second pop-up menu. You choose Changes because you want to trigger this email notification only if the status is changed. Click the Add (+) button at the right of the Metadata area to add additional criteria.

13 From the first pop-up menu, choose Status; from the second pop-up menu, choose Matches; and from the third pop-up menu, choose "Ready for Executive Producer Review."

14 Now your metadata subscription has been configured to look for a change in the Status field to "Ready for Executive Producer Review." Click Continue. We'll configure the email response in the next dialog.

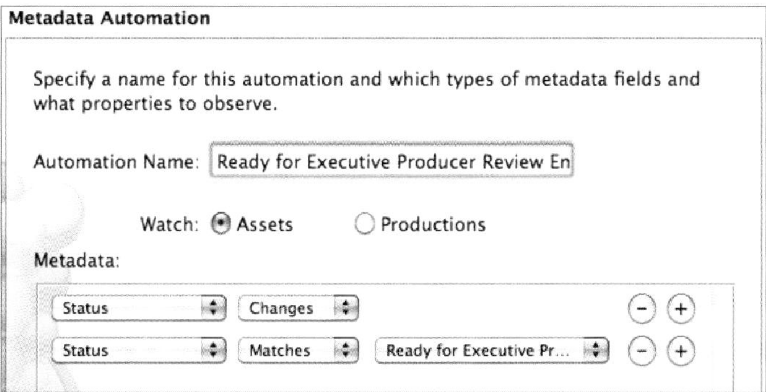

15 Click the Add (+) button at the bottom of the Responses dialog, and choose Email Response from the pop-up menu. For Name, enter *Email Response for Ready for Executive Producer Review*.

16 Complete the remaining fields as shown in the following image. Use the Insert Metadata Field pop-up menu for the bracketed ([ ]) text items.

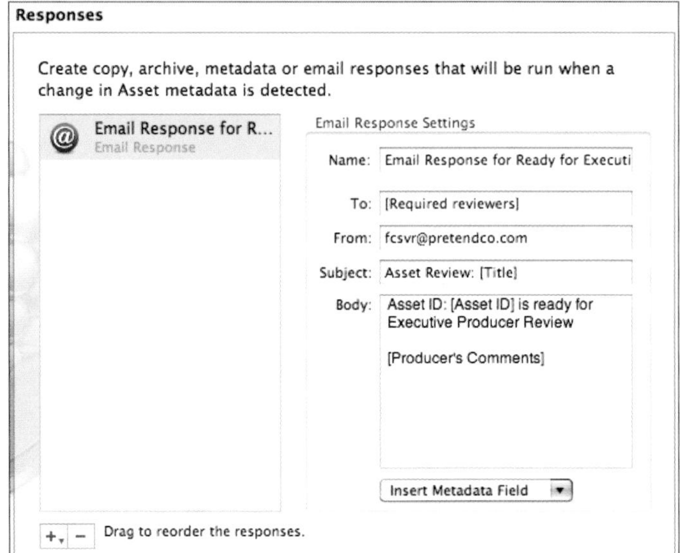

Your email response is now configured to send an email to the executive producer when "Ready for Executive Producer Review" is chosen from the Status pop-up menu.

17 Click Continue after you have reviewed what you entered to confirm that it matches the screen shot. Click Done on the next screen to finish adding your metadata subscription. After you click Done, you'll be sent back to the main Automations pane. Leave System Preferences open for now.

ADMIN ▼

Configuring Special Permission Sets

Now that you've configured your automations to send email notifications, you'll configure special permission sets for the producer and executive producer based on their user groups. This will allow them to see only the content that's pertinent to them. You'll modify one existing permission set, and then you'll create a new one for executive producers.

1 Bring the Administration window to the front of your desktop.

2 Click the Permission Set pane in the left menu, and double-click the reviewer permission set to open its settings.

3 Click Asset Filter in the left pane to open the Asset Filter pane. You modified the Asset Filter in an earlier lesson to customize your advanced search, and you can see here that the same metadata group is used in permission sets.

4 From the Status pop-up menu, choose Contains, and in the text-entry field, type *Ready for Producer Review*. Click Save Changes. This permission set will allow a user to see only assets that are marked as "Ready for Producer Review."

5 Back in the Permission Set pane of the Administration window, select reviewer and click the Duplicate button.

6 Double-click the duplicate of reviewer that is created, and change the name to *ExecProducer Review*.

7 Click Asset Filter in the list on the left, and change the Status pop-up menu to Contains. In the text-entry field, enter *Ready for Executive Producer Review*. Click Save Changes when you're finished.

   Now that you've modified your permission sets to suit your Review and Approve workflow, you have to apply them to the groups you created earlier in the lesson.

8 Bring the Final Cut Server preferences pane back to the front, and click the Group Permissions tab at the top. Notice that right now you have only one permission set in use for the admin user.

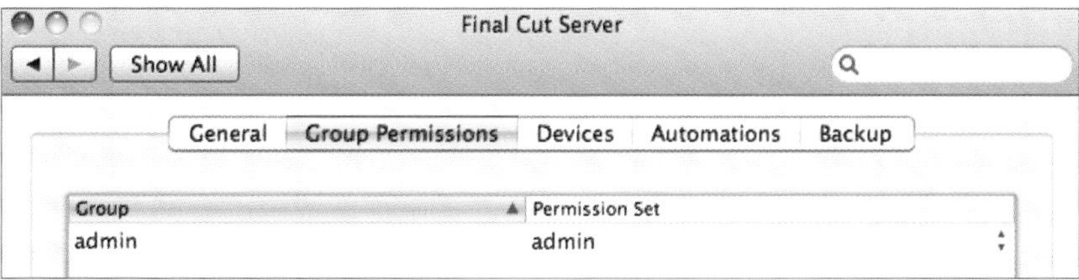

9 Click the Add (+) button in the lower-left corner, and choose the execproducer group. This group will now show up under the admin group, but with the same permissions as admin. Under the Permission Set column for the execproducer group, choose ExecProducer Review from the pop-up menu.

10 Click the Add (+) button in the lower-left corner, and choose the producer group. Under Permission Set for the producer group, choose reviewer from the pop-up menu.

11 Click the Add (+) button in the lower-left corner, and choose the editor group. Under Permission Set for the editors group, choose editor from the pop-up menu.

Now that you've applied the proper permission sets to your producer, executive producer, and editor users, they will be able to see only the assets in Final Cut Server that apply to their permission set. This prevents groups of users from seeing content that might confuse them during the Review and Approve process.

USER ▼

Walking Through a Custom Review and Approve Workflow

In the previous exercises, you modified the default Review and Approve template to better fit your review workflow. Now that you have everything configured, you'll step through the workflow as each person and highlight how this would look in a day-to-day Review and Approve workflow. Of course, your workflow will likely be different, but this exercise should give you a starting point for customizing Final Cut Server to fit your workflow.

You begin by logging in as each individual user to show which assets each user is able to see. This highlights the use of permission sets to restrict which assets users are able to interact with. After that, you log in as the editor, make some comments about an individual asset, and then change the status to Ready for Producer Review. This sends an email notification to the producer user with the Title, the Asset ID, and the comments that the editor left.

You then log in as the producer user, and note which assets he or she has access to. Because of the permission set applied to the Producer group, the producer user is able to see only assets marked as Ready for Producer Review. As the producer user, you leave comments about the asset in the Producers Comments metadata field, and then change the Status to

"Ready for Executive Producer Review." This sends an email notification to the executive producer user with the Title, the Asset ID, and the comments the producer left.

You then log in as the executive producer user, and note which assets he or she has access to. Because of the permission set applied to the ExecProducer group, the executive producer user is able to see only assets marked as Ready for Executive Producer Review. You then leave comments as the executive producer.

The next logical step in this workflow is delivering a finished asset. You'll cover this step in Lesson 9.

NOTE ▶ If you are not set up with an SMTP server, the email responses will not be sent. If so, you may switch users simply at the client application level.

1 To begin, log in to Mac OS X as the editor. You will need to install the client application. Refer to Lesson 1 for installation instructions.

2 In the client application, do a search for *Knight*, and double-click the asset to bring up the asset info window. Click the Review & Approve button at the top.

3 Enter some comments in the "Editor's comments" metadata field on the Review & Approve pane, such as *Here is my first attempt at the shot*. From the Status pop-up menu, choose Ready for Producer Review.

4 From the "Required reviewers" pop-up menu, choose Producer; and click Save Changes at the bottom of the window. This triggers the first step in your Review and Approve workflow by sending an email notification to the producer with the Title, Asset ID, and Editor's comments.

5 Log out and log back in with the producer user name and password. You'll need to download the client application again, as outlined in Lesson 1.

6 Open Mail, and configure it per your SMTP settings (available from your network administrator). Note the email that was automatically sent from Final Cut Server when the editor chose Ready for Producer Review from the Status pop-up menu. Also note that the editor's comments were inserted directly into the email.

NOTE ▶ If SMTP is not available, the email will not be sent/received.

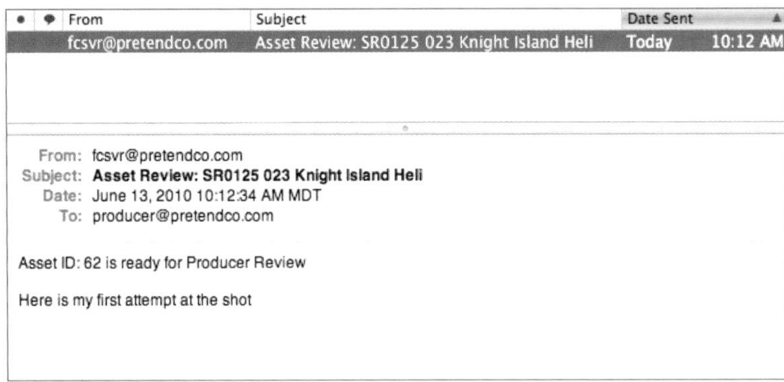

7 Log in to the client application as the producer user. Do a search without any keywords, and now you should see one asset.

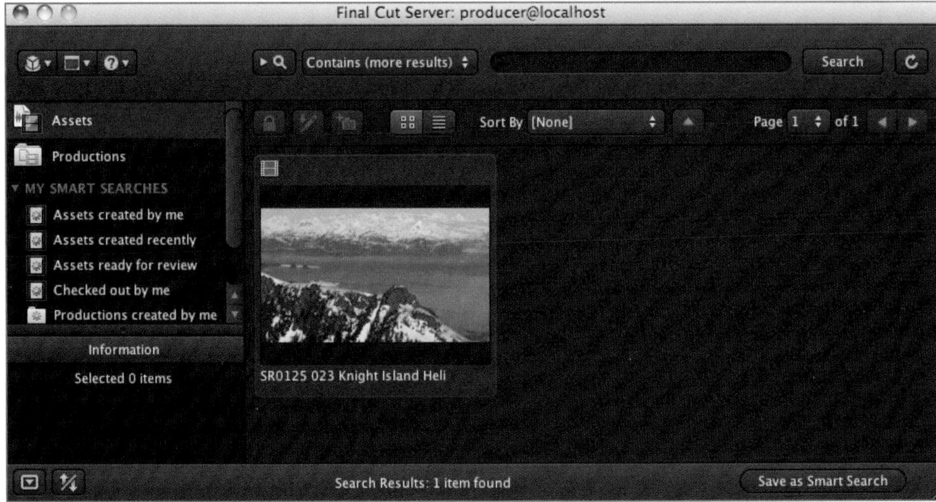

8 Double-click the asset **SR0125 023 Knight Island Heli** and navigate to the Review & Approve pane after the asset info window opens.

9 Enter some comments in the Producers Comments metadata field, such as *I think this shot is excellent, approved!* From the Status pop-up menu, choose Ready for Executive Producer Review; from the "Required reviewers" pop-up menu, choose Executive Producer; and click Save Changes. This will trigger your second automation to send an email to the executive producer.

10 Log out, then log in as the executive producer user with the executive producer credentials. Open Mail and configure the SMTP settings per your network/mail administrator. Notice that a new email has been sent and has the producer's comments from the asset.

NOTE ▶ If SMTP is not available, the email will not be sent/received.

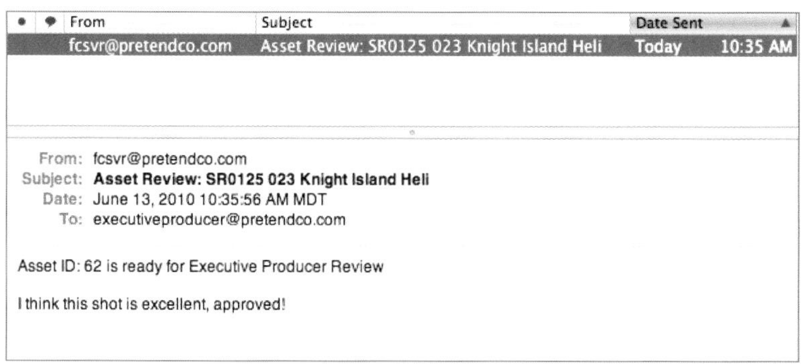

11 Re-download the client application as the executive producer user, as outlined in Lesson 1.

12 Log in to the client application as the executive producer, and note that you are now able to see the **SR0125 023 Knight Island Heli** asset. Again, this is because Ready for Executive Producer Review was chosen from the Status pop-up menu, and the permission set that's applied to the ExecProducer group allows only those users to see assets that have the correct status.

13 Double-click the asset, then click the Review & Approve button at the top.

14 Enter some comments in the Executive Producers Comments metadata field, such as *I agree with the Producer, approved!* From the Status pop-up menu, choose Approved. Click Save Changes, and close the asset info window.

Right now, changing the Status pop-up menu to Approved will not trigger any automations (you haven't set any up for Status equals Approved), but in the next lesson you will cover delivery of your finished assets.

ADMIN ▼

Using Annotations for Review and Approve

On top of the built-in Review and Approve metadata template, Final Cut Server also ships with an annotation feature that will allow you to make notes on individual assets based on timecode.

You can mark your ins and outs in the Annotations window and add those as comments. These comments become searchable by other users, just like any other metadata.

With Final Cut Server 1.5, you can also export these comments as XML. You'll want to output your annotations if you are integrating with a third-party CMS system, or if you have written your own web-based review and approve tool. One example of such a tool is on the Apple Final Cut Server website (www.apple.com/finalcutserver) under Resources, Web-Based Review and Approve tool.

You'll make individual comments as the editor and the producer users to show how a Review and Approve workflow would look inside the Annotations window. Then, you'll configure a new subscription to output XML with the comments and associate that with your Status pop-up menu. You'll also create a device to output this XML file into.

1 Log in to Mac OS X as the administrator. Open System Preferences, click the Final Cut Server icon, and authenticate.

2 Click the Devices button at the top of the Preferences pane, and click the Add (+) button in the bottom-left corner to make a new device (where you will be outputting your annotations as XML). Choose Local for the Device Type and click Continue.

3 For Device Name, enter *XML Out*, and then navigate to the FCSvr folder. Click New Folder, name the folder *XML Out*, and then click Create. Choose that folder, and when you're finished, click Continue.

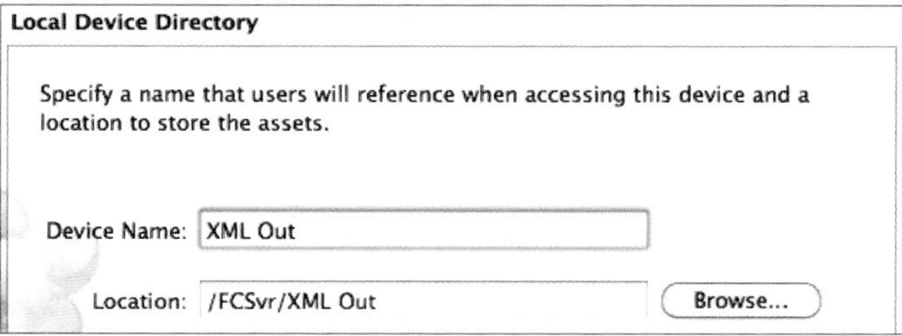

4 Leave the "Enable as an Archive Device" checkbox deselected and click Continue.

5 You will also not be scanning this device, as you are using it as output for your XML, so leave the scan settings deselected and click Continue.

6 There will not be any transcoding associated with this device, either, so you can leave the transcode settings at No Conversion. Click Continue.

7 When you're finished, click Done, and then quit System Preferences.

This new local device is now ready to receive your XML; you just need to configure the new XML responses.

8 Quit the client application if you have it open; you've made a change in the Final Cut Server preferences (added a device) that will not show up in the client application until it is reopened.

NOTE ► Because the client application is written in Java, when you change the metadata template it is necessary to redraw the Java window. This requires either opening up a new workspace or closing the client application and reopening it.

9 Log in to the client application as the administrator user. Open the Administration window, click Response in the list on the left, and click the Create button to make a new response.

10 In the Response window, choose Write XML from the Response Action pop-up menu. For Name and Description, enter *Annotations Export as XML.*

11 Click Write XML in the list on the left, and from the Destination pop-up menu, choose the XML Out device you created earlier. (If you don't see it, restart the client application.)

12 Select "Use ID for filename," and then click Save Changes.

The write XML response you've created will write out the annotations as XML into the XML Out device you created earlier. Now that you've made the response action, you need to create a trigger from a metadata subscription to actually call your write XML response. You'll create a metadata subscription to trigger your write XML response when you choose Accepted from the Status pop-up menu.

NOTE ► The reason you should select the "Use ID for filename" checkbox in a write XML subscription is to prevent assets with the same title from overwriting each other. If you have two assets with the title "Sunday" and they both have a write XML response associated with them, and they are both triggered, the latter XML file will overwrite the first XML file.

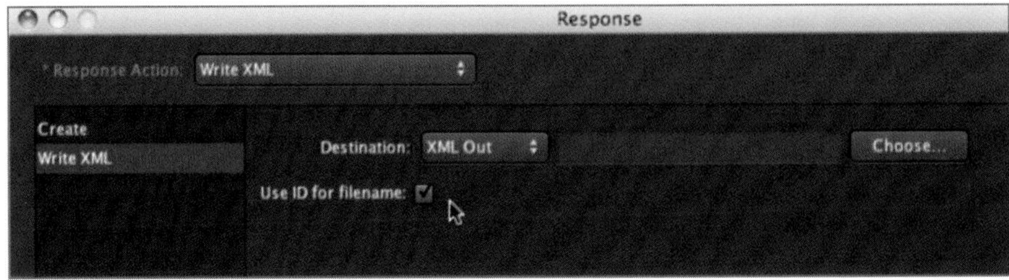

Earlier, you made metadata subscriptions through the Preferences pane, which gives only the option of associating a copy, archive, email, or delete response with each subscription. As you may have noticed in the Response Action pop-up menu, there are many more types of responses.

Now that you have made your write XML response through the Administration window, let's create a new metadata subscription from there as well, and associate your write XML response with it.

13 In the Administration window, click Subscription in the list on the left, and then click Create to make a new subscription.

14 From the "Subscribe to" pop-up menu, choose Asset (you are subscribing to a change in asset metadata).

NOTE ▶ You'll look at subscribing to jobs in Lesson 9, which talks about delivery. The other option is subscribing to a change in production metadata and subscribing to a system event.

15 In the Name field, enter *Export Annotations on Status Accepted.* Select the Enabled checkbox to turn on the subscription; and in the Description field, enter *Export Annotations on Status Accepted.* Leave Event Priority Modifier set to Normal (you can set the priority Higher or Lower depending on how many subscriptions you have going off at once).

16 For Event Type Filter, choose Modified, as you are looking for a change in metadata. From Response List at the bottom, choose your "Annotations Export as XML" response from the Available list, and click Add to move it to the Selected list.

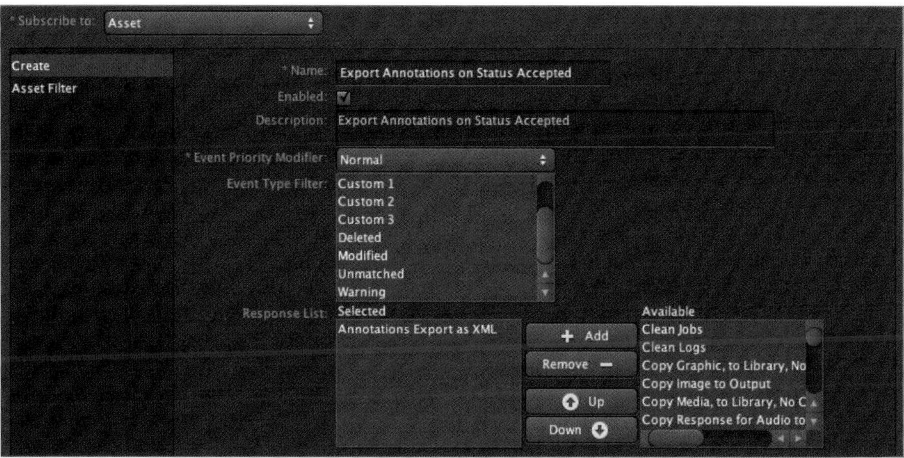

17 Click Asset Filter in the list on the left. From the Status pop-up menu, choose Equals; in the text-entry field, enter *Approved*, and select the "Trigger if changed" checkbox. Click Save Changes when you're finished.

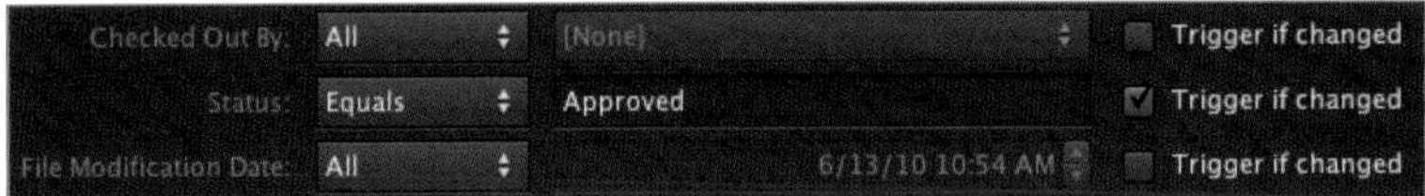

NOTE ▶ If you don't select "Trigger if changed" when creating a subscription through the Administration window, you'll likely run into a looping issue. Selecting "Trigger if changed" has the same behavior as selecting Changes through the Automation Assistant in the Preferences pane. Your response will fire only when there is a change in that metadata. Otherwise, if you don't select the checkbox, the subscription will fire off every time it finds a match, and this will likely cause an infinite loop that can be fixed only by stopping Final Cut Server.

Now your subscription is configured to output annotations from an asset that has the Status pop-up menu changed to Approved. The annotations will be exported as an XML file to your XML Out device.

USER ▼

Walking Through a Sample Review and Approve Workflow

Now that you have everything configured, let's take a step-by-step walk through a sample Review and Approve workflow with an editor and a producer, have them make some annotations, and then have them exported as XML.

1 Quit the client application, and log back in as the editor user.

2 Double-click the **SR1020 015 Glacier Island-v** asset to bring up the asset info window. Click the Annotate button to open the Annotations window.

3 Use the progress bar to scrub through the movie, mark an In point and an Out point using the tools on the right side of the viewer, and enter a note in the Annotation field, such as *Added some speed changes here*. Click Add Annotation at the bottom.

Note that your annotation is added in the right pane, with your user name and timestamp and the In and Out points. You can jog to the specific points in the clip by clicking the arrow buttons next to your timecode from the annotation.

4 Close the Annotations window viewer, and click the Review & Approve button at the top of the asset info window. From the Status pop-up menu, choose Ready for Producer Review; from the "Required reviewers" pop-up menu, choose Producer; and click Save Changes.

As occurred earlier in this lesson, this will send an email notification to the producer user that the asset is ready for review, but note that this email does not include the annotations.

5 Restart the client application, and log in as the producer user.

6 Double-click the **SR1020 015 Glacier Island-v** asset. Click the Annotate button to open the Annotations window. Note that you can see the annotation that the editor left earlier.

7 Scrub to any point in the Viewer, mark an In point and an Out point, and add the note *Speed change looks great, approved!* in the Annotation field. Click Add Annotation to add it to the annotations on the right.

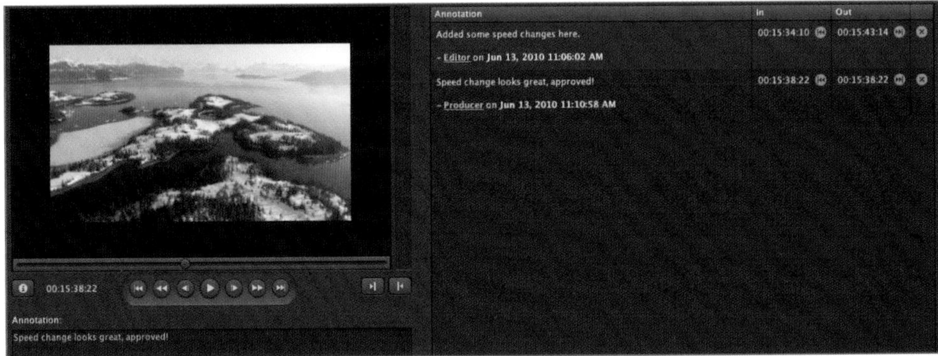

8 Close the Annotations window, and click the Review & Approve button at the top of the asset info window. From the Status pop-up menu, choose Approved, and click Save Changes.

This will trigger your write XML response and will export all the metadata associated with the asset and the annotations as an XML file to your XML Out device.

9 Open a Finder window, and navigate to the FCSvr/XML Out folder, where you configured the local device earlier. Notice that there is now an XML file inside that directory, titled with the asset ID of the asset that triggered the response (the filename may vary, depending on the order of steps you took in the book, as the asset IDs may be different).

10 Right-click the XML file and choose Open With > Other from the shortcut menu. In the Choose Application window, select to open it in TextEdit. Note the highlighted portion in the following screen shot; these are your annotations exported as XML. The rest of the information is all the metadata that was associated with this asset.

```xml
    <mdValue fieldName="Last Accessed" dataType="dateTime">2010-06-13 17:12:21+0</mdValue>
    <mdValue fieldName="Video Elements" dataType="string">video</mdValue>
    <annotations>
     <annotation>
      <mdValue fieldName="User id" dataType="integer">2</mdValue>
      <mdValue fieldName="Full name" dataType="string">Editor</mdValue>
      <mdValue fieldName="Out" dataType="timecode">00:15:43:14/(24000,1001)</mdValue>
      <mdValue fieldName="Annotation" dataType="string">Added some speed changes here.</mdValue>
      <mdValue fieldName="In" dataType="timecode">00:15:34:10/(24000,1001)</mdValue>
      <mdValue fieldName="Timestamp" dataType="dateTime">2010-06-13 17:06:02+0</mdValue>
      <mdValue fieldName="Name" dataType="string">editor</mdValue>
     </annotation>
     <annotation>
      <mdValue fieldName="User id" dataType="integer">3</mdValue>
      <mdValue fieldName="Full name" dataType="string">Producer</mdValue>
      <mdValue fieldName="Out" dataType="timecode">00:15:38:22/(24000,1001)</mdValue>
      <mdValue fieldName="Annotation" dataType="string">Speed change looks great, approved!</mdValue>
      <mdValue fieldName="In" dataType="timecode">00:15:38:22/(24000,1001)</mdValue>
      <mdValue fieldName="Timestamp" dataType="dateTime">2010-06-13 17:10:58+0</mdValue>
      <mdValue fieldName="Name" dataType="string">producer</mdValue>
     </annotation>
    </annotations>
  </metadata>
 </entity>
```


Integrating this XML with a third-party system is outside the scope of this book, but getting it exported is the first step. The next step would be parsing this information, and then getting it into a third-party system.

Lesson Review

1. What is the function of permission sets?
2. Where is Final Cut Server's Review and Approve functionality located?
3. True or false: You can edit a lookup through a System Preferences pane.
4. How can you export annotations from Final Cut Server?
5. Why do you want to select the "Trigger if changed" checkbox when you create a metadata subscription through the Administration window?

Answers

1. Permission sets allow you to restrict your users' access to assets and what they are allowed to do with those assets and associated metadata.
2. The Review and Approve workflow metadata fields, lookups, and automations are located in the pane accessed by clicking the Review & Approve button at the top right of all asset's info windows.
3. False. The Administration window is the only place you can create, modify, or delete lookups.
4. Annotations can be exported from Final Cut Server as XML by using a write XML response associated with a metadata subscription.
5. If you don't select "Trigger if changed," you could trigger an endless loop because the value being subscribed to is always being satisfied.

Goals

Use watchers to distribute assets

Use subscriptions to distribute assets

Lesson **9**

Distributing Finished Media

After you've discovered your acquired assets and gone through the editorial and Review and Approve process, you're now ready to deliver your media. In Final Cut Server, there are two main methods for distributing finished media: using watchers and metadata subscriptions.

In Lesson 4, you learned how to use watchers to add files to the Final Cut Server catalog as assets. The methodology is the same in this lesson, except in this case you don't add the files to Final Cut Server as assets; you transcode and deliver them to their intermediate or final destinations. Watch folders allow you to automate these copy and transcode responses by stacking responses on top of each other. For example, you can have a watcher that creates deliverables for Apple TV, iPhone, or a web streaming clip, and then sends an email notification after they're delivered.

ADMIN ▼

Configuring Settings for Publishing Finished Media

Before you figure out whether your workflow is best suited to use watchers or metadata subscriptions, you must first decide where you're going to publish your content. Final Cut Server supports local file systems (local folders), network file systems (AFP, SMB, NFS, FTP), and Xsan file systems for devices. Since you might not have the necessary infrastructure to use each of the file system types, in this lesson you'll set up dummy network and Xsan devices as local folders. This will allow you to learn the procedures for publishing to these destinations without having the necessary equipment. The procedures for adding network and Xsan devices are outlined in the *Final Cut Server Setup Guide*.

In this lesson, you'll create three devices: FTP, Local, and Xsan. For each of those devices you'll make three destination folders, each for a specific content type: H.264 for Apple TV, H.264 for iPhone and iPod, and MPEG-4. In the next exercise, you'll use watchers to publish and transcode content to these destination devices, and then in the exercise after that, you'll do the same exercise using metadata subscriptions.

1 Log in as administrator to your system, open the Finder, navigate to the FCSvr directory that you created earlier, and make a new folder inside that directory called *Destinations*.

2 Inside the Destinations directory, make three new folders: *FTP*, *Xsan*, and *Local*. These are the dummy devices where you'll publish content via both watchers and metadata subscriptions.

3 Inside each of the folders you just made, make three new subfolders: *H264 for AppleTV*, *H264 for iPhone*, and *MPEG4*.

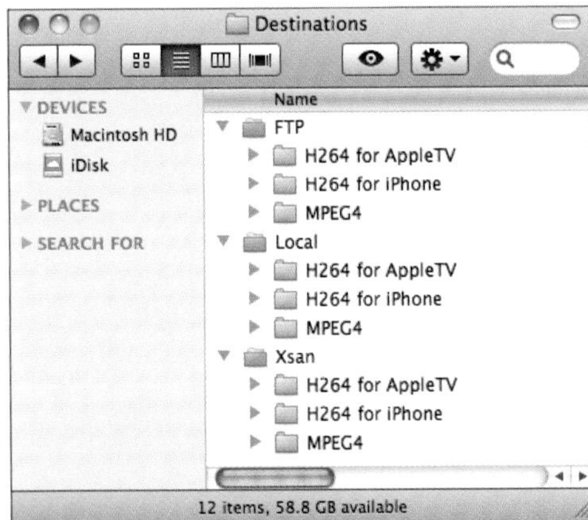

4 Open System Preferences, click the Final Cut Server icon, and then authenticate.

5 Click the Devices button and then the Add (+) button in the bottom left. Select Local in the Device Setup Assistant and click Continue.

NOTE ▶ If you were configuring a true FTP device, you would have clicked on the Network button instead and entered the relevant FTP info (login name, password, hostname). Likewise, when adding a true Xsan device, you'd need a true Xsan volume, and Final Cut Server would have automatically configured edit-in-place for you. In this exercise, you will choose Local for all three devices.

6 For Device Name, enter *FTP*, navigate to the /FCSvr/Destinations/FTP location, and click Continue.

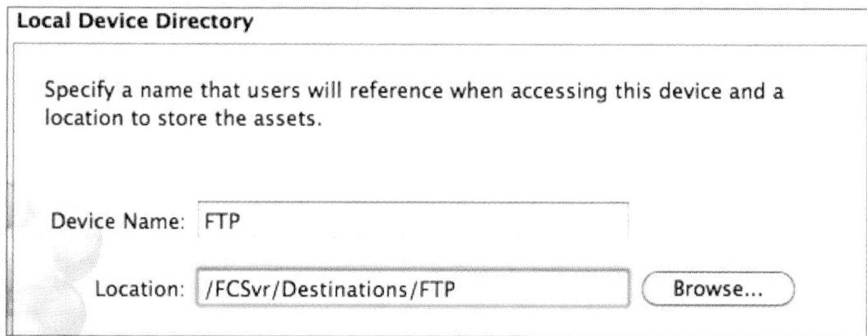

7 Don't select the "Enable as an Archive Device" checkbox; you won't archive anything to this device, because you're pushing finished content to a delivery destination. Click Continue to move on to the Scan Settings dialog. You won't scan this location either, as again, you're only pushing content for delivery to external systems. Click Continue. The Transcode Settings dialog appears.

Remember how you made three folder types under your destination devices at the beginning of the exercise? This is where you configure the transcode settings.

8 Scroll through the list and enable the following: H.264 for AppleTV, H.264 for iPod Video and iPhone 640x480, and MPEG-4. Checking these transcode settings allows you to automatically transcode an asset to the specified format through a watcher or through a metadata subscription. Click Continue when finished.

9 At the final dialog of the Device Setup Assistant, verify that all of your settings are correct, and then click Done to finish adding your new device with its configured transcode settings.

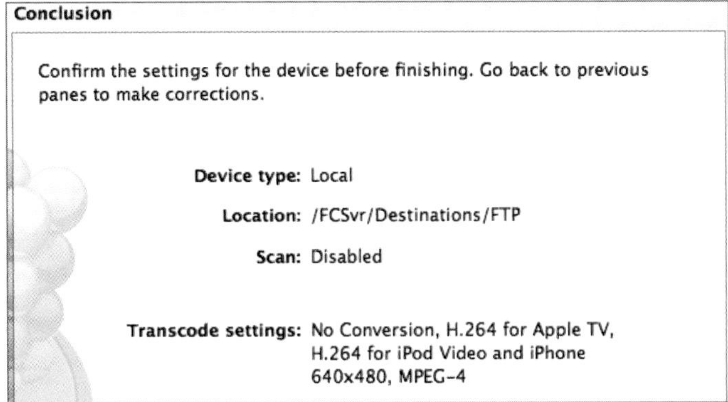

10 Repeat steps 5–9 for the two other destination folders you made earlier, Xsan and Local. Give the files names that correspond to the folder in use and choose the same transcode settings. When finished, the three devices should look like this:

FTP	/FCSvr/Destinations/FTP	Local
Xsan	/FCSvr/Destinations/Xsan	Local
Local	/FCSvr/Destinations/Local	Local

Configuring Job Email Notifications

Now that you've finished configuring the destinations where you're going to publish finished material, let's configure a useful email notification for transcoded and delivered content. In this exercise you'll also choose default settings for Maximum Running Jobs (the maximum number of transcodes that happen concurrently), Retry Count (the number of times to retry a failed job until Final Cut Server gives up), and Retry Timeout (the amount of time to pause between retrying the failed job).

Sometimes jobs fail to publish. This can be caused by any number of factors, but the most likely is network interference and a destination server going through a crash/power cycle. Tweaking the default settings can prevent jobs from failing permanently, especially if you know the power cycle times of your equipment. The email notification is sent when the job ultimately fails and not during any of the retry counts. The notification is nice, because the only other way to know that a job fails is to constantly monitor the Search All Jobs window (which is difficult if these transcodes or publishes happen during off hours) or to notice that the file never shows up on the destination device. We highly recommend setting up an email list or email alias that references all the relevant people that need to be notified. Sending these notifications to one person isn't always efficient (since your single recipient might not be there to receive the email notification).

1 If necessary, return to the Final Cut Server pane in System Preferences.

2 In the General pane, note the default setting for Max Running Jobs. It's currently set to 10, which is a good value for running a contained Final Cut Server system that does the transcoding locally. If you were integrating Final Cut Server with an Apple Qmaster cluster, you could safely bump up this number based on the number of nodes (your mileage may vary depending on system specs) in increments of five jobs until you reach a satisfactory threshold.

3 For Retry Count, enter *5*; and for Retry Timeout, enter *60*. These values tell Final Cut Server to retry a failed job five times, pausing 60 seconds between each try (most modern servers should power cycle in 5 minutes). You can modify these values based on your infrastructure, taking into account the time it takes your destination servers to go through a power cycle (one of the more likely causes of a failed job). When finished, quit System Preferences.

4 Open the client application and log in as the administrator. Open the Administration window from the Server pop-up menu.

5 Choose Response from the menu on the left, and click the Create button to make a new response. Choose Email from the Response Action pop-up menu. For Name, enter *Failed Job Email Notification*, and for description, enter *Email notification sent when a job fails.*

6 Choose Email in the menu on the left and enter the following values (they will differ based on your email infrastructure): For To, enter the email address you want to have the notifications sent to (such as *sysadmin@pretendco.com*); for Sender, enter *fcsvr@pretendco.com*; and for Subject, enter *Job [Job ID] Failed*. For Message, enter the following:

Job [Job ID] [Title] needs attention. Please login ASAP and attempt to troubleshoot.

The error that occurred was:

"[Error]."

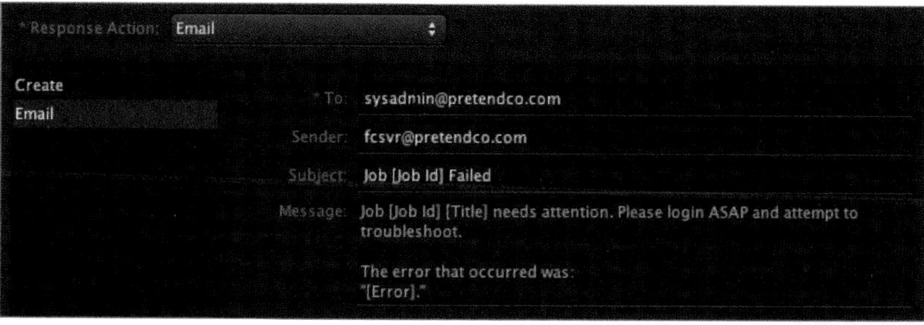

NOTE ► As you learned in Lesson 8, you can automatically insert asset metadata into email notifications. You can do the same thing with metadata from job logs. Notice in the figure that you insert the Job ID, the title of the job, and the actual error that's generated. Having all this information directly in the email saves the system administrator time as they will know what is wrong with the job instantly. They can do all of this without having to log in to the client application, select the individual job, and then look at the job logs.

7 When you're finished, click Save Changes to save your new email response.

8 Back at the Administration window, choose Subscription from the menu on the left. Click the Create button to make a new subscription.

9 From the Subscribe To pop-up menu, choose Job (you are looking for a change in job metadata now, not asset metadata). For Name, enter *Failed Job Notification*, and select the Enable checkbox. For Description, enter *Failed Job Notification*; for Event Priority Modifier, choose Normal from the pop-up; and for Event Type Filter, choose Modified (you are looking for a change in metadata). From the list of Available responses, choose the Failed Job Email Notification response you created earlier in this exercise and click Add to move it to the list of Selected responses on the left.

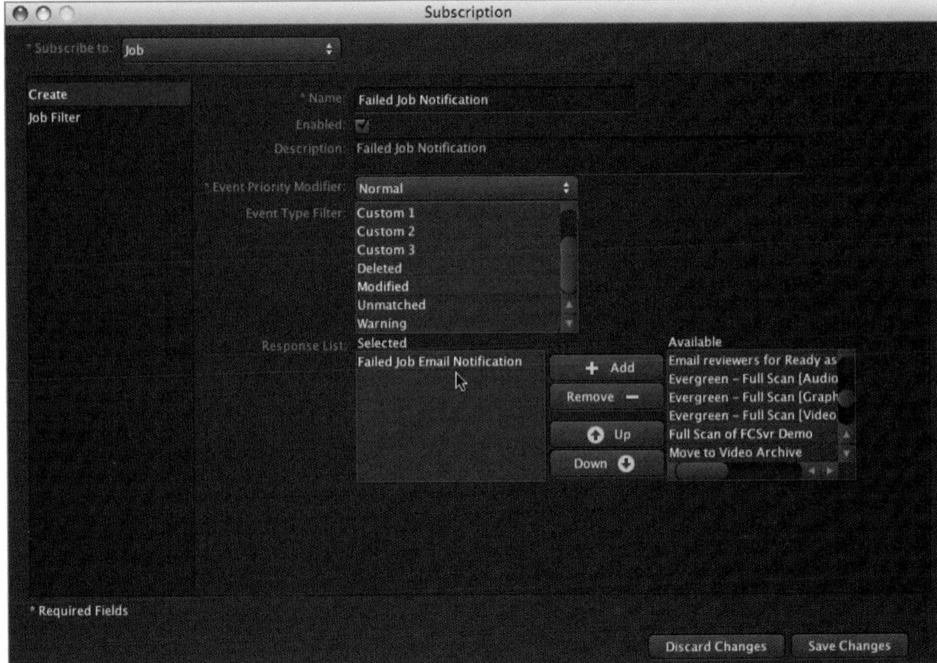

10 Choose Job Filter from the menu on the left, and choose Equals from the Status pop-up menu. Enter *FAIL* in the text-entry field next to the Status pop-up (you are looking for jobs that have a status of failed), and select "Trigger if changed." Click Save Changes when finished.

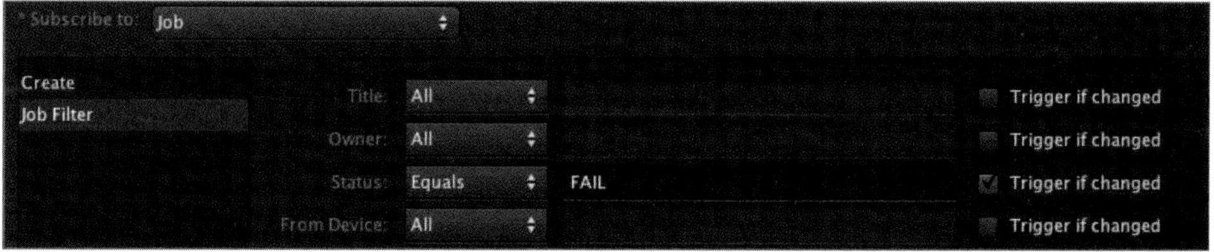

Now when a job fails, it automatically sends out the email notification you created. This is because you set up a subscription looking for a change in the job status to Failed, and this change triggers your email being sent.

ADMIN ▼

Configuring Watchers for Delivery

In the previous exercise, you configured your destination folders, devices, and transcode settings. Now you'll set up your watchers and the copy responses that do the actual transcoding and pushing of the content to your destination devices. Watchers are useful for delivering finished content; there's no need to log directly in to Final Cut Server. An editor can export a finished sequence directly from Final Cut Pro to a Final Cut Server watcher. The editor's work is now complete. Final Cut Server takes over the task of transcoding the media into the proper formats and pushing that transcoded media to its destination location.

In this exercise, you'll configure three watchers. Each one will transcode the media dropped into it into the three settings you configured earlier (H264 for AppleTV, H264 for iPhone, and MPEG-4) and push it to the destination device associated with the watcher.

1 Log in to your system as administrator. Open the Finder and navigate to the FCSvr directory you created at the root of the drive. Inside the FCSvr directory, double-click Watchers. This is where your watch folders will reside.

NOTE ▶ In a real-world scenario, the watchers will reside on the server and be shared to the client computers (or directly mounted in the case of a Xsan).

2 Inside the Watchers folder, create three new directories: *FTP*, *Local*, and *Xsan* (Graphic and Media come with the default template). This is where you drop your content to be transcoded and delivered.

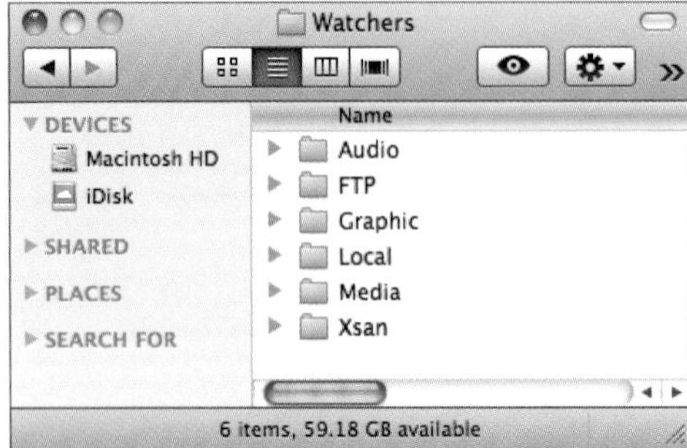

3 Open System Preferences, click the Final Cut Server icon, and authenticate. Click the Automations button, and click the Add (+) button in the bottom left to add a new automation.

4 In the Automation Setup Assistant, select File System Watcher and click Continue. In the next dialog, for Automation Name, enter *Deliver Finished Media to FTP*.

5 From the Device pop-up menu, choose Watchers (this is the location where your watch folders reside, not the location to which you'll publish content); select the Watch Subfolder checkbox; click Browse and choose FTP as the folder you'll be watching.

6 For Filter, click the Add (+) button on the bottom left and choose *.MOV (you want only movies, not any other files). When finished, click Continue.

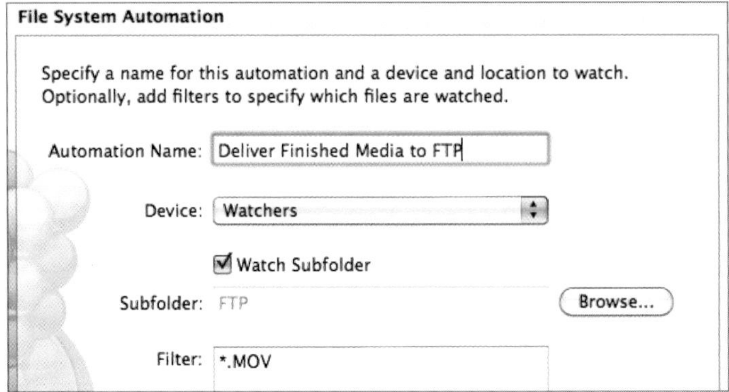

In the next dialog you'll set up your copy responses, which apply your transcode settings, and then copy the content to its final destination.

7 In the Responses dialog, click the Add (+) button at the bottom left and choose Copy from the pop-up menu. For Name, enter *Publish iPhone to FTP*; for Destination Device, choose FTP from the pop-up. Select the Destination Subfolder checkbox (you want to publish each type of media to its respective subfolder), click Browse, choose "H264 for iPhone," and then click OK.

8 From the "Transcode source to" pop-up, choose "H.264 for iPod Video and iPhone 640x480." Leave the Metadata Set pop-up at None.

NOTE ▶ You leave this setting at None because you're not adding this file to the asset catalog, just pushing it to its final destination.

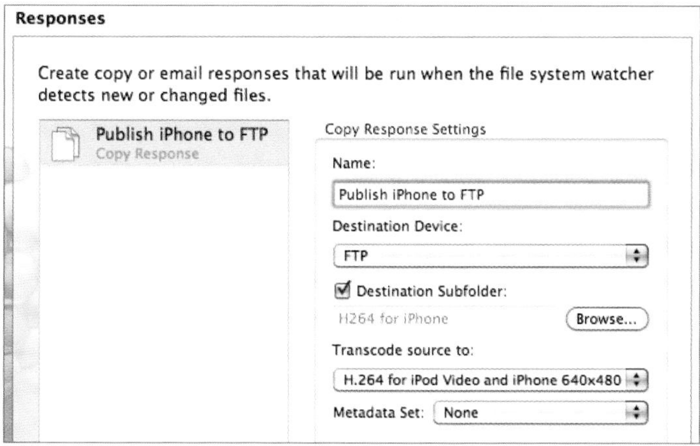

9 Still in the Responses dialog, click the Add (+) button at the bottom left and choose Copy from the pop-up menu.

10 For Name, enter *Publish AppleTV to FTP*; for Destination Device, choose FTP from the pop-up. Select the Destination Subfolder checkbox, click Browse, choose "H264 for AppleTV," and then click OK.

11 From the "Transcode source to" pop-up, choose "H.264 for Apple TV." Leave the Metadata Set pop-up at None.

12 In the Responses dialog, click the Add (+) at the bottom left and choose Copy from the pop-up menu.

13 For Name, enter *Publish MPEG-4 to FTP*; for Destination Device, choose FTP from the pop-up. Select the Destination Subfolder checkbox, click Browse, choose MPEG4, and then click OK.

14 From the "Transcode source to" pop-up, choose MPEG-4. Leave the Metadata Set pop-up at None.

15 In the Responses dialog, click the Add (+) button at the bottom left, and choose Delete from the pop-up. Make sure the delete response is under all the copy responses (you can drag to reorder the items).

NOTE ▸ The reason you add a delete response is so that the files do not fill up space in the watcher directory. If you want to save the original file, you add another copy response (with No Conversion as the transcode setting) to push the source material into another internal device of Final Cut Server.

When finished, your watcher setup should look like the following screen shot. Click Continue when ready, and then click Done.

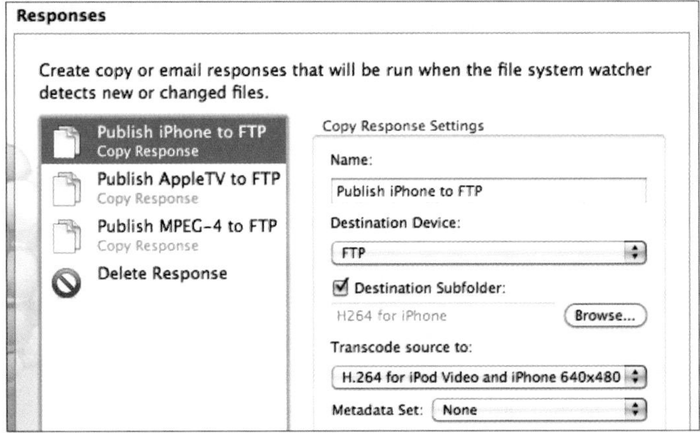

Next you'll add the other two watchers and their copy responses.

16 Back at the main Automations pane in Final Cut Server System Preferences, click the Add (+) button to add a new automation. Choose File System Watcher and click Continue. For Automation Name, enter *Deliver Finished Media to Local*. From the Device pop-up menu, choose Watchers (this is the location where your Watch folders reside, not the location to which you'll publish content). Select the Watch Subfolder checkbox, click Browse, and choose Local. For Filter, click the Add (+) button and choose *.MOV. When finished, click Continue.

17 In the Responses dialog, click the Add (+) button at the bottom left and choose Copy from the pop-up menu. For Name, enter *Publish iPhone to Local.* For Destination Device, choose Local from the pop-up. Select the Destination Subfolder checkbox, click Browse, choose "H264 for iPhone," and then click OK. From the "Transcode source to" pop-up, choose "H.264 for iPod Video and iPhone 640x480." Leave the Metadata Set pop-up at None.

18 In the Responses dialog, click the Add (+) button at the bottom left and choose Copy from the pop-up menu. For Name, enter *Publish AppleTV to Local.* For Destination Device, choose Local from the pop-up. Select the Destination Subfolder checkbox, click Browse, choose "H264 for AppleTV," and then click OK. From the "Transcode source to" pop-up, choose "H.264 for Apple TV." Leave the Metadata Set pop-up at None.

19 In the Responses dialog, click the Add (+) button at the bottom left and choose Copy from the pop-up menu. For Name, enter *Publish MPEG-4 to Local.* For Destination Device, choose Local from the pop-up. Select the Destination Subfolder checkbox, click Browse, choose MPEG4, and then click OK. From the "Transcode source to" pop-up, choose MPEG-4. Leave the Metadata Set pop-up at None.

20 In the Responses dialog, click the Add (+) button at the bottom left and choose Delete. Make sure it's under all of the copy responses. When finished, click Continue, and then Done.

Now you'll add your final watcher.

21 Back at the main Automations pane, click Add (+) to add a new automation. Choose File System Watcher and click Continue. For Automation Name, enter *Deliver Finished Media to Xsan*. From the Device pop-up menu, choose Watchers (this is the location

where your watchers reside, not the location to which you'll publish content). Select the Watch Subfolder checkbox, click Browse, and choose Xsan. For Filter, click the Add (+) button on the bottom left and choose *.MOV. When finished, click Continue.

22 In the Responses dialog, click the Add (+) button at the bottom left and choose Copy from the pop-up menu. For Name, enter *Publish iPhone to Xsan*. For Destination Device, choose Xsan from the pop-up. Select the Destination Subfolder checkbox, click Browse, choose "H264 for iPhone," and then click OK. From the "Transcode source to" pop-up, choose "H.264 for iPod Video and iPhone 640x480." Leave the Metadata Set pop-up at None.

23 In the Responses dialog, click the Add (+) button at the bottom left and choose Copy from the pop-up menu. For Name, enter *Publish AppleTV to Xsan*. For Destination Device, choose Xsan from the pop-up. Select the Destination Subfolder checkbox, click Browse, choose "H264 for AppleTV," and then click OK. From the "Transcode source to" pop-up, choose "H.264 for Apple TV." Leave the Metadata Set pop-up at None.

24 In the Responses dialog, click the Add (+) button at the bottom left and choose Copy from the pop-up menu. For Name, enter *Publish MPEG-4 to Xsan*. For Destination Device, choose Xsan from the pop-up. Select the Destination Subfolder checkbox, click Browse, choose MPEG4, and then click OK. From the "Transcode source to" pop-up, choose MPEG-4. Leave the Metadata Set pop-up at None.

25 In the Responses dialog, click the Add (+) button at the bottom left and choose Delete. Make sure it's under all of the copy responses. When finished, click Continue, and then Done. You may quit System Preferences.

USER ▼

Copying Content from Watchers to Destination Locations

Now that you have configured your watchers for distributing your files, let's see them in action. You'll drop some sample content into each of your watchers and observe them copying the content to the destination locations you created earlier in the lesson.

1 In a new Finder window, open the Watcher Media folder from the FCS_Book_Files folder.

This is the content you'll use to demonstrate delivery via watch folders. You don't add it to the asset catalog as these are not source clips, but deliverables.

2 In another Finder window, navigate to the FCSvr directory and open the Watchers folder.

This is where the watchers you created reside, and you'll copy the sample content into these folders to trigger your publish responses.

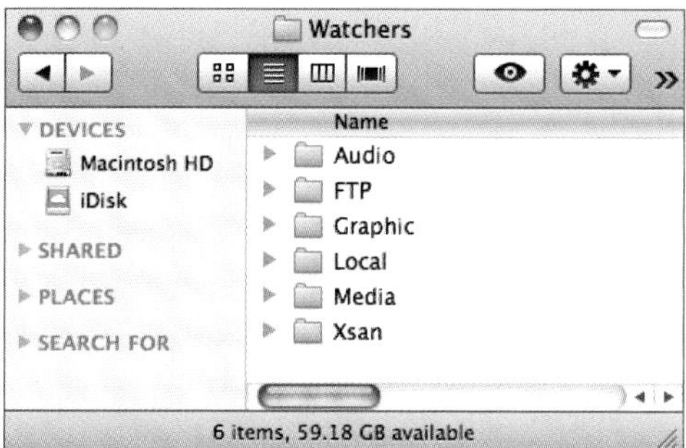

3 Copy the file **SR0134 003 The Needles Sealions** from the Watcher Media folder to the FTP folder in the Watchers directory. Since we made these watchers in the Preferences pane, they will have a default polling value of around 30 seconds.

4 Open the Search All Jobs window in the client application and observe the multiple copies and transcodes that happen. Depending on the system you've installed Final Cut Server on, these transcodes could take upward of 5 minutes.

NOTE ▸ When the transcodes have finished, note that the file that you dropped into the FTP folder is no longer there. This keeps the watcher directories clean.

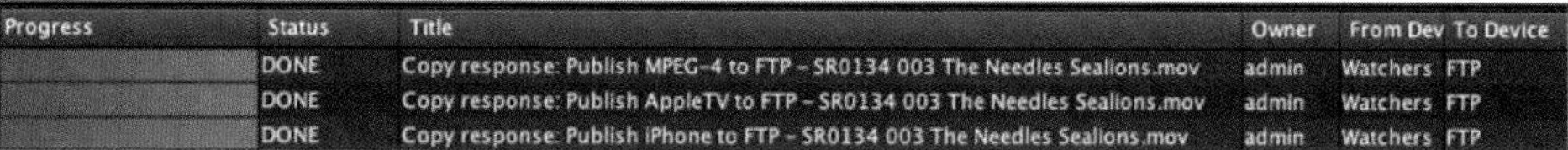

5 Navigate to the /FCSvr/Destinations/FTP directory and notice that the finished clips have been copied directly to these locations.

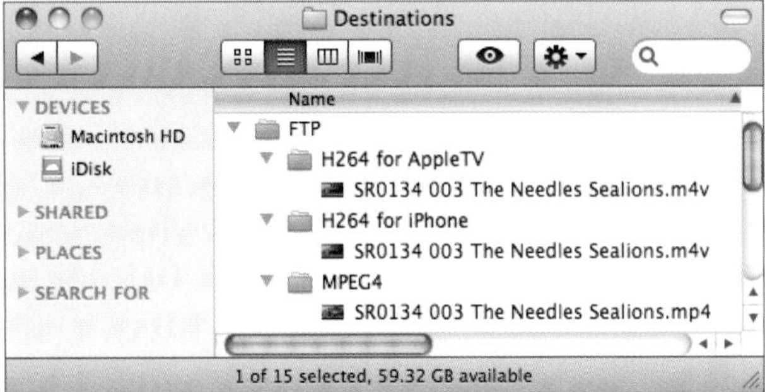

6 Highlight the **SR0134 003 The Needles Sealions** file in the MPEG4 folder in the FTP directory and press Command-I to bring up information about that file.

7 Click the disclosure triangle next to More Info to show the codec information about the file. Note that it has been transcoded from its original ProRes 422 (LT) format to MPEG-4. This was the direct result of the copy response you set up with the watcher, which automatically transcoded any file dropped inside it to MPEG-4.

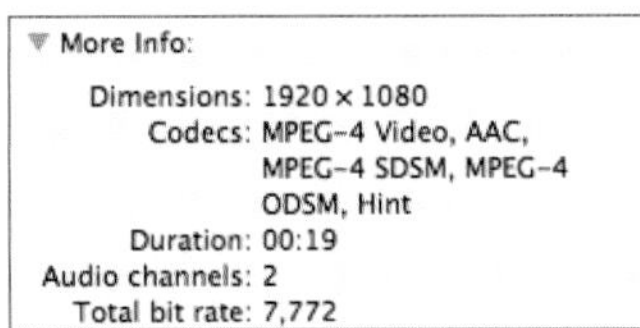

8 Copy the file **SR0185 007 Smith** from the Watcher Media folder to the Local folder in the Watchers directory. Open the Search All Jobs window in the client application and observe the multiple copies and transcodes that happen.

When the jobs have finished, navigate to the /FCSvr/Destinations/Local directory and notice that the finished clips have been copied directly to these locations.

9 Copy the file **SR0512 009 Iyoukeen** from the Watcher Media folder to the Xsan folder in the Watchers directory. Open the Search All Jobs window in the client application and observe the multiple copies and transcodes that happen.

When the jobs have finished, navigate to the /FCSvr/Destinations/Xsan directory and notice that the finished clips have been copied directly to these locations.

ADMIN ▼

Configuring Metadata Subscriptions for Delivery

In the previous exercises, you learned how to use watchers to transcode and distribute finished material. This happens outside the client application, while delivering via metadata subscriptions happens entirely inside the client application.

The benefit of using metadata subscriptions is that they're more flexible than watchers. If you need to deliver to multiple locations and multiple formats, you can find yourself having to deal with a lot of folders very quickly. You can also accidentally drop the files into the incorrect folder, which would not trigger any action or may trigger additional actions you didn't anticipate.

In this exercise, you'll configure Boolean metadata fields (yes/no checkboxes) to trigger the copy responses you made earlier, then create a new metadata group called Publish Destinations. First, you need to make your new metadata fields, then you associate those fields with the new metadata group you made. Next, you add these fields to the Asset Filter metadata group so they're available for metadata subscriptions. Finally, you associate the copy responses you made earlier with the Boolean fields through metadata subscriptions.

1 Open the Administration window of the client application.

2 In the left pane, choose Metadata Field and click the Create button. In the Metadata Field window that appears, for Name enter *iPhone FTP*, choose Boolean from the Data Type pop-up, enter *iPhone FTP* for Description, choose None from the Category pop-up, and leave the rest of the values at default. Click Save Changes when finished.

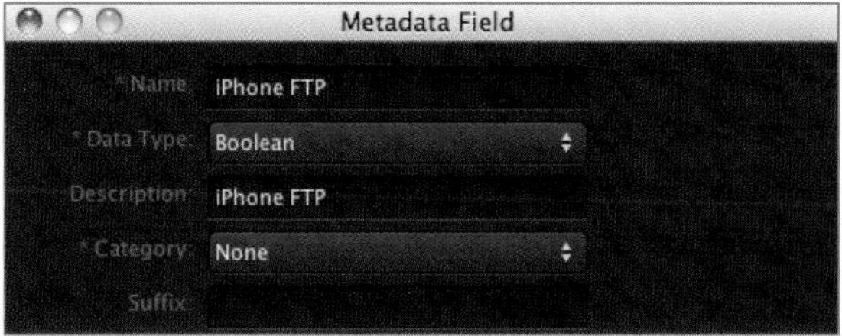

3 Back at the Metadata Field window in the Administration window, click the Create button to make a new field. For Name enter *iPhone Xsan*, choose Boolean from the

Data Type pop-up, enter *iPhone Xsan* for Description, choose None from the Category pop-up, and leave the rest of the values at default. Click Save Changes when finished.

4 Back at the Metadata Field in the Administration window, click the Create button to make a new field. For Name, enter *iPhone Local.* Choose Boolean from the Data Type pop-up, enter *iPhone Local* for Description, choose None from the Category pop-up, and leave the rest of the values at default. Click Save Changes when finished.

5 Back in the Administration window, choose Metadata Group in the menu on the left, and click the Create button in the top left of the Metadata Group area to make a new metadata group.

6 In the New Metadata Group window, for Name, enter *Publish Destinations.* From the Available list, highlight the iPhone FTP field you just created, and click Add to move it to the Selected list on the left. Do the same thing for the other two fields you created, iPhone Local and iPhone Xsan. When finished, you should have the three new metadata fields you created in steps 2–4 in the Fields Selected list on the left. By doing this, you're making the metadata fields you created available to the Publish Destinations metadata group that you're creating.

TIP ▶ The new metadata fields will appear at the bottom of the Available list.

7 In the Actions area, highlight the Create, View Details, and Edit Details actions on the Available list on the right, and click Add to move them to the Selected list on the left.

NOTE ▶ Adding these actions makes the metadata group available to Final Cut Server for creating new assets (Create), and for viewing (View Details) and editing (Edit Details) metadata from those assets.

8 Under the Metadata Sets area, highlight Media (Media Asset) in the Available list on the right, and click Add to move it to the Selected list on the left. You can ignore the Display Hints section at the bottom, as there's nothing to change there. Click Save Changes when finished.

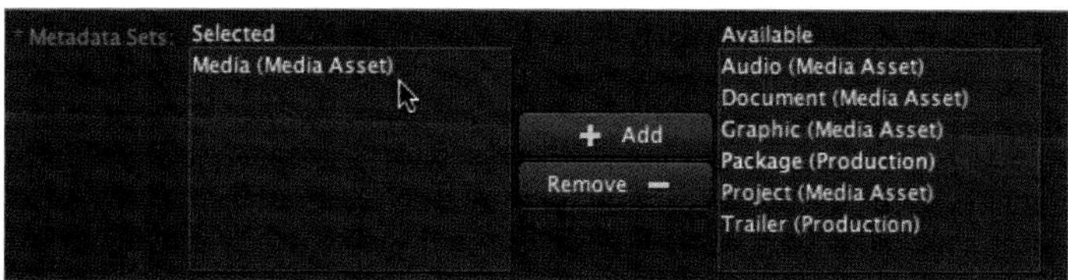

ADMIN ▼

Associating and Triggering Responses Using a Metadata Subscription

Now that you've created your fields and associated them in the Publish Destinations metadata group, you need to associate the copy responses you created earlier in the lesson with these fields using a metadata subscription. By using Boolean metadata fields, you'll be able to trigger your copy and transcode responses by having a user simply check the appropriate Boolean field. This allows automations to trigger existing assets in the catalog from within the client application, while automations associated with watchers are triggered outside the client application.

First, you must associate the new fields you made with the Asset Filter metadata group.

NOTE ▶ As you learned earlier, the Asset Filter group is used in advanced searches, in permission sets, and also with metadata subscriptions. Editing this group makes your custom fields available to those functions of Final Cut Server.

1 In the Administration window of the client application, choose Metadata Group in the left menu. In the search field on the right, enter *Asset Filter* and click Search. Double-click Asset Filter from the search-results list, which opens that metadata group for editing.

2 In the Fields area, in the Available list on the right, select the iPhone Local, iPhone Xsan, and iPhone FTP metadata fields, and click Add to move them to the Selected list on the left. To see your newly added fields in the Selected list, you need to scroll down. Click Save Changes when finished.

Now that you've associated your fields with the Asset Filter metadata group, you're able to trigger automations using metadata subscriptions. Let's set one up in the System Preferences pane.

3 Open System Preferences, click the Final Cut Server icon, and authenticate.

4 Click the Automations tab along the top, and then click the Add (+) button in the lower left to create a new automation. Click the Metadata Subscription button and click Continue.

5 For Automation name, enter *iPhone FTP Delivery Subscription*. Under Watch, leave Assets enabled, and under Metadata from the first pop-up choose iPhone FTP. From the second pop-up, choose Changes. Click the Add (+) button on the far right to add new criteria, and from the pop-up choose iPhone FTP. From the second pop-up, choose Matches, and from the last pop-up, choose True. Click Continue.

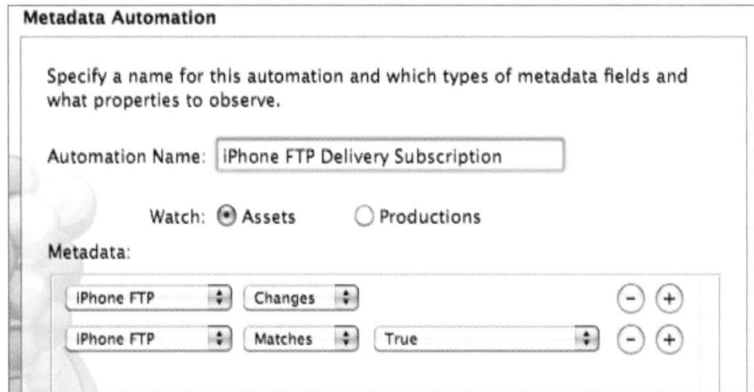

6 On the Responses pane, click Add (+) to add a new response and choose Copy Response. For Name, leave the default. For Destination Device, choose FTP, and select the Destination Subfolder checkbox. Then click Browse, select the "H.264 for iPhone" folder, and click OK. From the "Transcode source to" pull-down, choose "H.264 for iPod Video and iPhone 640x480," and leave Metadata Set at None. Click Continue, and then click Done. Quit System Preferences.

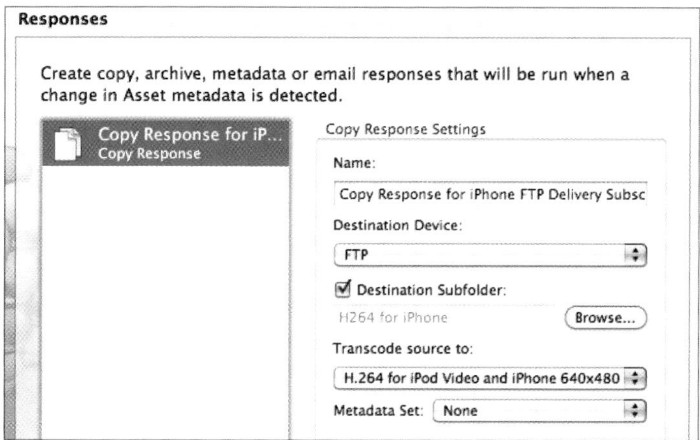

USER ▼

Using Metadata Subscription for Publishing

Now that you've learned how to properly configure a subscription based on custom metadata, next you'll trigger your subscription by selecting the iPhone FTP checkbox on one of your assets.

1 Launch the client application and log in as editor if the window is not already opened. (If you are still logged in as administrator, that's fine.)

2 Perform a search for *SR0125 023 Knight Island Heli* and double-click the asset that is returned. After the asset information window opens, click the Publish Destinations metadata group to expose the Boolean checkbox fields.

3 Select the iPhone FTP checkbox and click Save Changes. This will trigger the metadata subscription that you set up in the previous steps. You can monitor the job by launching the Search All Jobs window.

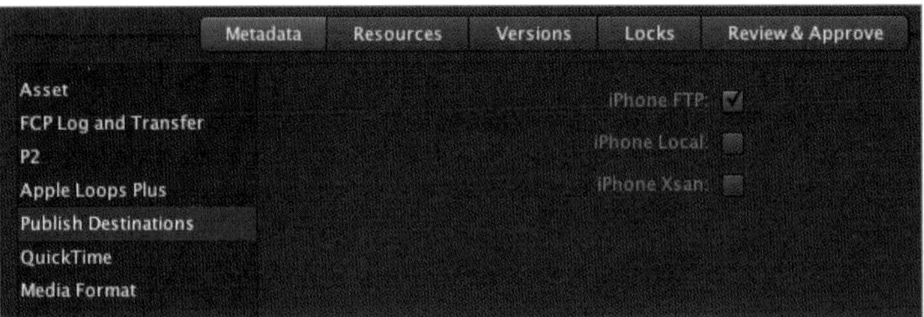

4 To verify that the job has completed correctly, you can navigate to the device location in the Finder, search for the completed job in the Jobs window, or you could log in as the administrator, launch the Administration window, and go to the Logs tab and search for your job.

NOTE ▶ A common practice for diligent administrators is to set up a notification email attached to a job subscription to monitor the status of jobs. You can subscribe to all Jobs, Failed Jobs, or Completed Jobs, depending on how much email you want to receive.

Lesson Review

1. What delivery methods can you use within Final Cut Server?
2. To add custom subscription metadata fields, what metadata group do you need to edit?
3. True or false: You can use the System Preferences pane to associate custom metadata fields with subscriptions.
4. What formats does Final Cut Server support for delivery?

Answers

1. You can use watch folder automations or metadata subscriptions to automate the delivery of your finished media through Final Cut Server.
2. The Asset Filter metadata group.

3. False. You need to use the Administration window in the client application to associate custom metadata fields with subscriptions.
4. Final Cut Server supports any of the formats that Compressor supports (including plug-ins).

10

Goals

Creating Archive devices

Adding Archive Status to the Asset filter metadata group

Archiving and restoring assets

Automating the archiving process

Lesson **10**

Archiving

The archival process is the last step in your digital asset workflow. The main point of archiving is to move the primary representation of an asset in Final Cut Server to a predetermined location. The most likely motivation for moving the primary representation is disk space constraints. This location can be anything that presents itself as a file system and Final Cut Server has access to, such as a nearline spinning-disk storage that is fibre-attached, an external FireWire drive, or an HSM (Hierarchical Storage Manager) tape-based archival system. In an HSM system, an entire robotic tape library is abstracted by a simple file system interface, which works quite well in a Final Cut Server environment.

As long as Final Cut Server can read and write files to the file system, that device can be used for archiving within Final Cut Server. The key advantage to using Final Cut Server to manage your archival efforts is that Final Cut Server keeps the metadata and proxy files online. These archival actions are automated and can be triggered by users without any intervention from administrators, freeing up the admins to handle more important tasks. Assets are still available for updating metadata, and you can still view the proxies and mark notes on them using the Annotations window.

In real-life scenarios, the location that you would archive to would be some type of redundant storage (HSM, nearline spinning-disk, and so forth). In this lesson, we'll demonstrate how to archive to local archive devices, but you can easily apply the same workflow and configuration to other types of archive devices as long as they present themselves as some sort of mountable file system.

In Lesson 8, you learned how to use metadata subscriptions with the built-in Review & Approve metadata template to trigger email notifications. You'll use the same methodology in this lesson to trigger archive and restore responses, instead of email notifications.

Before you begin the exercises, here is a definition that will help with the terminology used in this lesson:

- An *archive device* is a special device type in Final Cut Server that can be specially flagged to hold primary representations of assets in the catalog.

This term appears throughout this lesson to show you how to access and use these asset representations.

ADMIN ▼

Creating an Archive Device

To Final Cut Server, an archive device is a folder to which Final Cut Server will archive the primary representations of assets. The media is moved from its original location to the folder that is designated as the Archive device. On that Archive device, the directory paths of the primary representation will be replicated from the source device with the first directory being the Device ID of the source device. The primary representation is then placed in the same file path on the archive device.

In this exercise you'll create three new archive devices (Video Archive for video files, Audio Archive for audio files, and Image Archive for still images) using the System Preferences pane, and you'll see how the archival process has been achieved.

1. Log in to your system as administrator and navigate to the /FCSvr directory. Create a new folder at that level named *Archive*. In that new folder, create three subfolders: *Video Archive*, *Image Archive*, and *Audio Archive.*

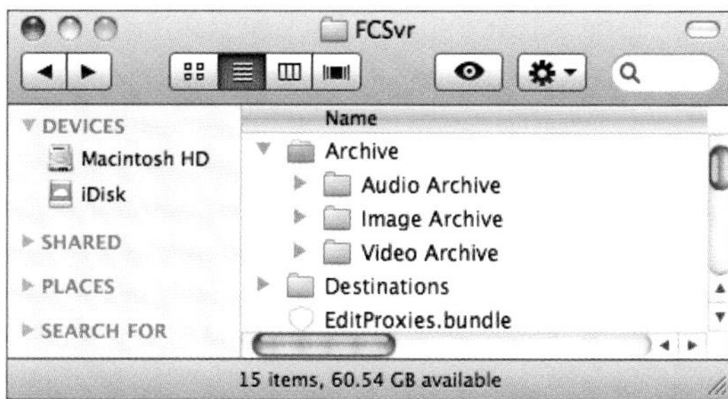

2 Open System Preferences, click the Final Cut Server preference pane, and authenticate as an administrator.

3 Click the Devices tab and then click the Add (+) button. In the Device Setup Assistant, choose Local for Device Type and click Continue.

4 For Device Name, enter *Video Archive*, and then browse to the FCSvr/Archive/Video Archive location and click Continue.

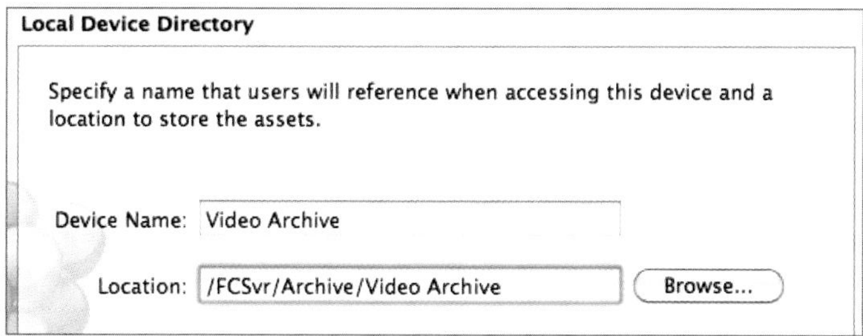

5 Select the "Enable as an Archive Device" option. Click Continue.

This adds the newly created device to the list of devices to which Final Cut Server is allowed to archive assets.

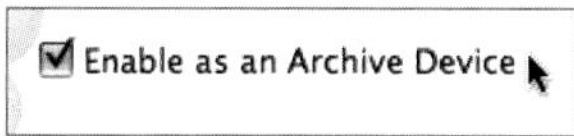

6 On the final summary screen, click Done.

Notice that the Device Setup Assistant automatically skipped the Scanning Assistant and Transcode Settings screens because you selected this device as an archive device.

7 Repeat steps 3 through 6 to create two additional local archive devices: *Audio Archive* and *Image Archive*. Select the relevant location that corresponds to the name of the device you are creating (for example, the Image Archive folder in /FCSvr/Archive/Image Archive).

8 Quit System Preferences when done.

Now that you've created three archive devices, you'll walk through the process of archiving and restoring your assets, and also learn some best practices when it comes to quickly retrieving and identifying archived media.

USER ▼

Archiving Assets

During the archiving process, Final Cut Server moves the primary representation of the selected assets to the specified archive device. The proxies and metadata information remain online, so the proxies are still viewable and the metadata can be searched and updated if needed. Keeping the proxies online also allows you to continue to use the Annotations window even though the primary representation has been moved offline.

In this exercise, you'll select a number of assets to archive. You'll utilize the client application to archive assets to their relevant archive device and show what this actually looks like in the file system.

1 Open the client application by double-clicking the Final Cut Server icon on your desktop, and log in as the administrator.

2 Perform a search without any keywords to show all available assets in the catalog.

3 Right-click the **1000Hz** asset (an audio file), and choose Archive to > Audio Archive from the shortcut menu.

This moves the primary representation associated with the selected asset to the Audio archive device.

NOTE ▶ The Archive shortcut menu will appear only if you have specified devices as Archive. If you have not set up any archive devices, you won't see the Archive option in the shortcut menu.

Also, you won't see the shortcut menu if you right-click FCP projects because they are not supported for archival purposes. You should utilize version control to create redundant versions of your Final Cut Pro project files.

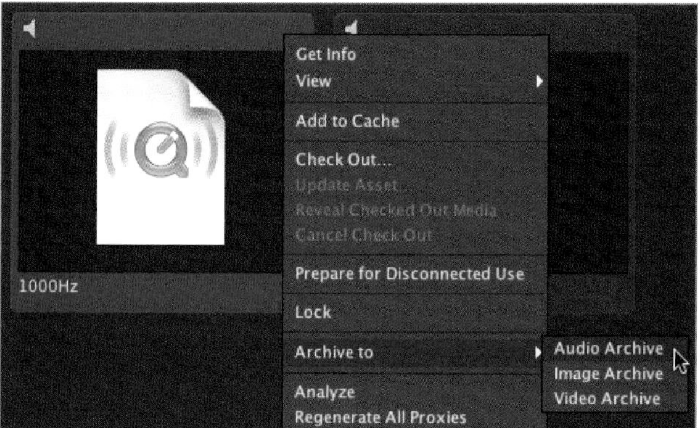

Notice that the archive icon (a filing cabinet) appears on the asset's Thumbnails view as a visual identifier that the asset has been archived.

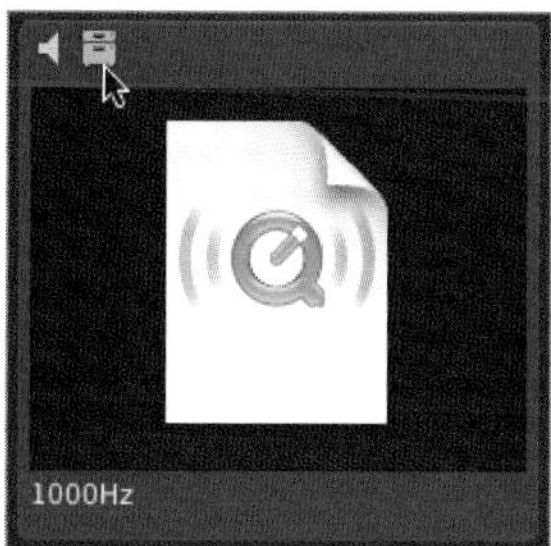

4 Right-click the **SR0489 004 Jackpot** asset (a video file) and choose Archive to > Video Archive from the shortcut menu.

This moves the high-resolution video file to the Video Archive folder you created earlier; however, the clip proxy is still available.

5 Click the Video Clip icon (filmstrip) on the **SR0489 004 Jackpot** asset.

The clip proxy for the asset will be cached and then opened into QuickTime Player.

6 Quit QuickTime Player and return to the client application.

7 Double-click the **SR0489 004 Jackpot** asset's thumbnail. In the asset info window, click the Resources tab.

The low-resolution proxies and metadata stay online and in the catalog.

NOTE ▶ The primary representation of **SR0489 004 Jackpot** is no longer listed; instead it's stored as the Archive Copy on the Video Archive device.

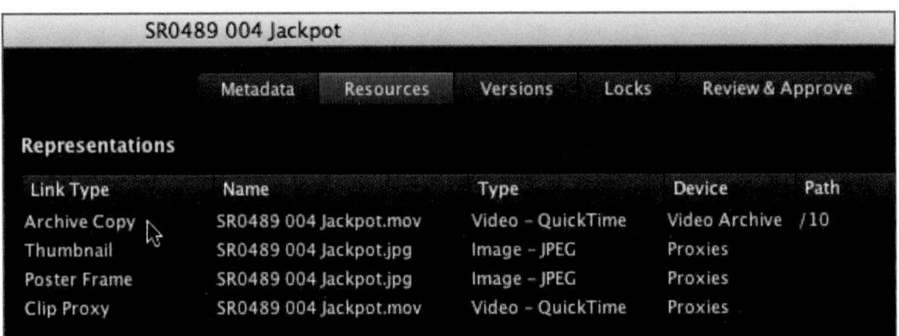

8 Back in the asset pane, right-click the **Evergreen Logo (Keyable)** asset (an image file) and choose Archive to > Image Archive.

This moves the high-resolution image from its original location to the Image Archive folder you created earlier in the lesson.

Now that you've successfully archived some material, let's take a look at the file system to see where the files have gone.

9 Open the Finder and navigate to the FCSvr/Archive directory at the root of your hard disk. Expand each archive directory and the links under them, and you'll see that during the archive process Final Cut Server has maintained the files and the file system structure as outlined earlier in the chapter.

NOTE ▶ My directories start with the number 13, as the device the media was archived from was the 13th device I've made on my system. Your Device ID is going to be lower (or higher) depending on the order in which you have gone through the exercises.

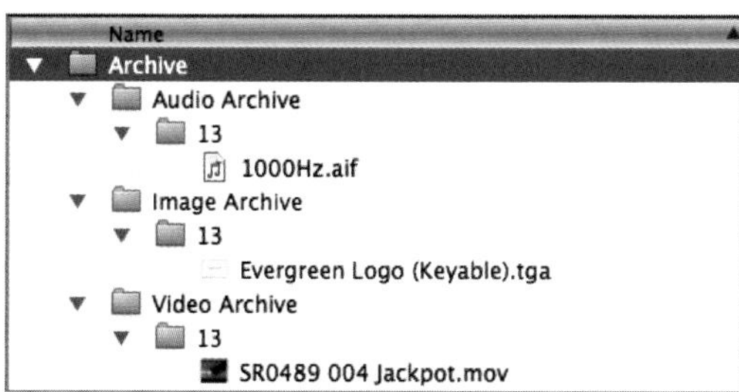

NOTE ▶ You've been individually archiving assets to highlight that you can have multiple archive devices, but you can also select multiple assets and archive them to the same device.

Now that you've archived some assets, you'll learn some quick ways to find them and utilize the content.

ADMIN ▼

Searching for Archived Assets

Archived assets will still show up in search results like any other asset type, but they will have an archive icon as a reminder that their primary representations are now offline.

Now that you've archived multiple assets, you'll look at the process of searching for those assets and updating metadata. During this process, you'll add the Archive Status metadata field—which isn't attached to a metadata set by default—to the Asset Filter metadata group. This field will help you locate all of your offline assets much more quickly.

1 In the client application, open the Administration window from the Server menu pop-up.

2 Choose Metadata Group from the pane on the left and do a search for *Asset Filter*. Double-click Asset Filter in the pane below to open the Asset Filter metadata group for editing.

3 From the list of available metadata fields on the right, select Archive Status and click Add to move it to the list of Selected fields on the left. Doing this will make the Archive Status metadata field available for your users through the Advanced Search pane. Click Save Changes when finished.

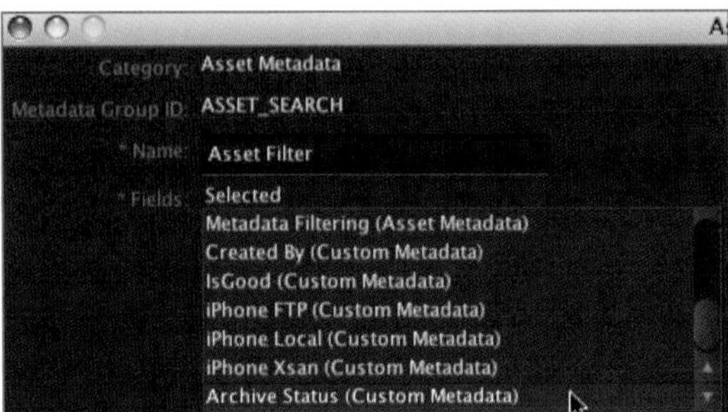

4 Quit the client application and then reopen it and log in as the administrator.

This allows you to see the new changes you made to the Advanced Search field.

5 Click the disclosure triangle next to the magnifying glass to open the Advanced Search pane. For Archive Status, choose Equals from the first pop-up menu.

Notice that the Archive Status field is now a filterable option with three choices:

- Offline—For assets that have been archived
- Online—For assets that are in an online status
- Unknown—For assets that have their archived version in an unknown state (for example, archiving to tape and then putting the tape on the shelf)

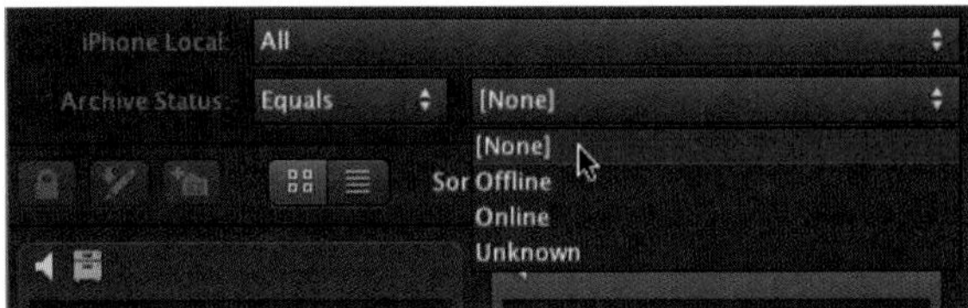

6 Choose Offline from the second pop-up menu, and then click Search. Notice that the returned results are the assets you archived earlier in the lesson.

7 Click the "Save as Smart Search for All Users" button and name the search *Archived Material.*

Now, any time any of your users log in, they will automatically be able to search for all archived assets with one click.

USER ▼

Restoring Archived Assets

Sometime during the lifecycle of your media, it's quite likely that you'll need to call back an asset from the archive. Final Cut Server manages this process by moving the archived primary representation back to the original location. This operation is handled entirely through the client application.

1 Open the client application and log in as the editor.

2 On the left panel in the client application, click the Archived Material smart search to quickly reveal all of the archived assets in the catalog, including the three assets you archived.

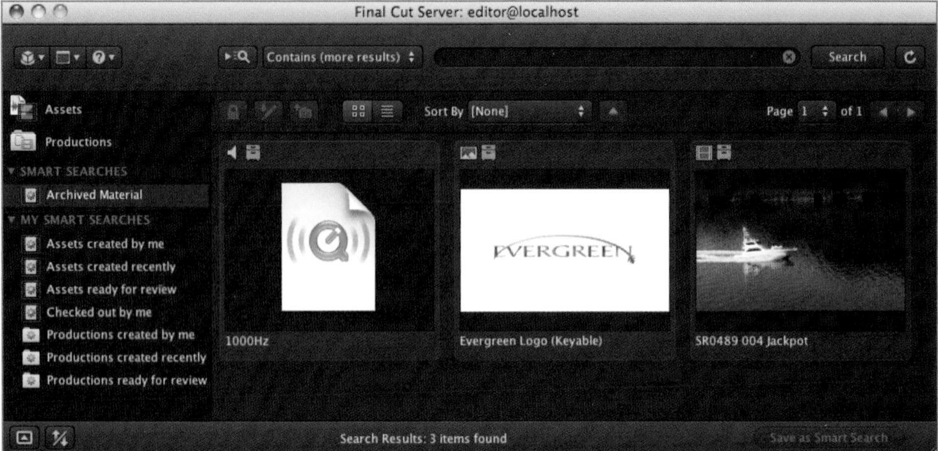

3 Click the archive icon of the **Evergreen Logo (Keyable)** asset, or right-click and choose Restore from the shortcut menu.

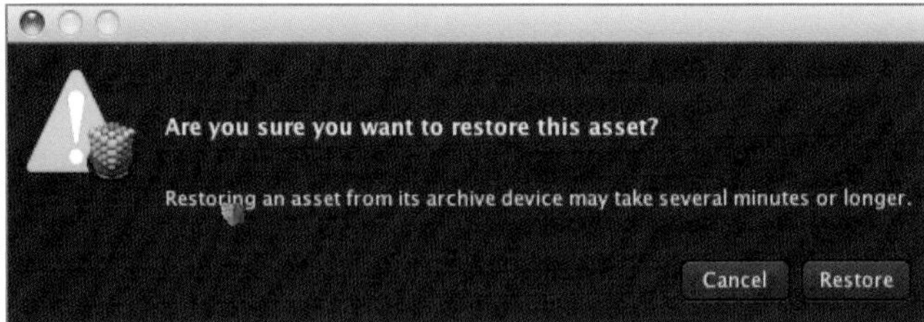

4 Click Restore to move the primary representation from the archive device location back to its original device and file system location. If your archive device or device you are restoring to is a network device (or if you have really big files), this process can take a long time.

5 Restore the other two assets back to their primary locations as well.

6 Open the Finder and navigate to the FCSvr/Archive location where you observed your archived content being moved to in the previous section.

Note that the archived media is now gone. It has been moved out of these folders and back to its original devices and locations.

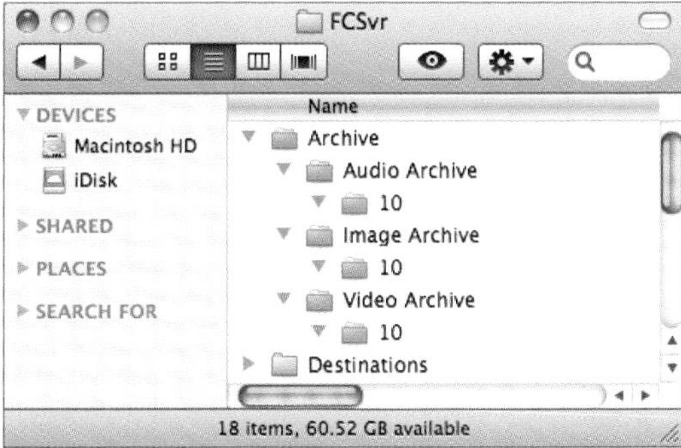

7 Back in the client application, double-click the **SR0489 004 Jackpot** asset and click the Resources tab to view information about its representations.

Note that the archive copy is now gone, and the primary representation is listed again on its original device and location.

What you've learned in this exercise is the manual way to archive and restore assets. This process can be automated using metadata subscriptions, which you'll set up next.

ADMIN ▼

Automating Archiving

Now that you've learned the fundamentals of how Final Cut Server handles the archiving process, let's look at automating the process using metadata subscriptions and responses. You can easily integrate the archive step into your workflow directly after delivering your finished assets. Adding an archive response after your copy response moves the high-resolution content offline after the finished assets have been delivered to their destinations.

In this section you'll create an archive response, which you'll use to move your high-resolution movie files offline after they've been transcoded and delivered. Remember the metadata subscription you made in Lesson 8 to deliver iPhone material? You're going to modify it to include an archiving step after the material has been transcoded and delivered.

1 Relaunch the client application and log in as the administrator. Open the Administration window from the Server menu pop-up.

2 Click Response from the panel on the left, and click the Create button to make a new response. From the Response Action pop-up menu, choose "Move to Archive." Enter *Move to Video Archive* for Name, and enter *Move primary representation to Video Archive* for Description.

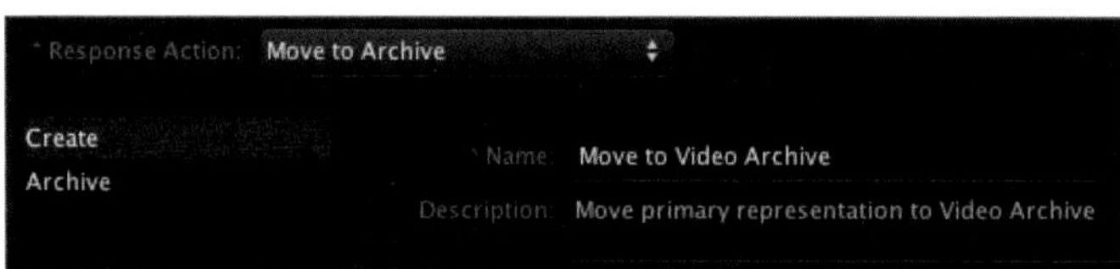

3 Select Archive in the left panel and choose the Video Archive device from the Archive Device pop-up menu. Click Save Changes.

4 Back in the main Administration window, choose Subscription from the pane on the left and double-click "Export Annotations on Status Accepted" to edit the subscription.

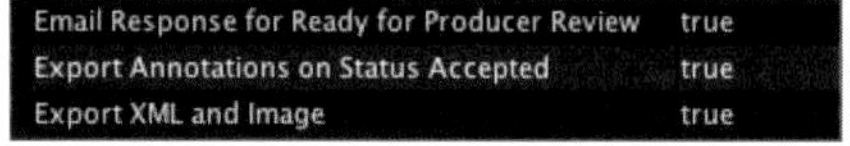

5 In the Response List pane, from the list of available devices on the right, choose "Move to Video Archive" and click Add to move it to the list of Selected responses on the left. It should appear under the existing Annotations Export as XML. If it

doesn't, use the Down button to move it beneath the original response. Click Save Changes when finished.

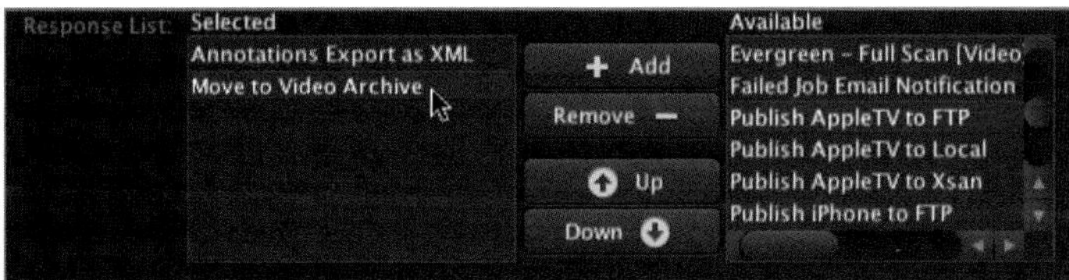

6 Back in the main client application window, double-click the **SR1015 002 Copper River Delta-v** asset. Click the Review & Approve tab along the top of the Metadata window, choose Approved from the Status pop-up menu, and click Save Changes.

   This triggers the metadata subscription you triggered earlier in the book, and also your automated archiving process.

7 Open the Search All Jobs window from the main client application pop-up menu and note that changing the metadata has triggered your subscription and your asset has been archived.

8 Back in the main client application window, click the Refresh button in the top right to refresh your search results. Note that the **SR1015 002 Copper River Delta-v** asset now also has the archive icon as a visual indicator that it has been archived.

You could take this subscription a step further by adding an email notification, or a script response to publish information to a third-party archival database.

Setting Up the Automated Restore Process

Now that you've successfully automated the archiving of your assets after they've been transcoded and delivered, you'll create a process to automate restoring assets from the archive.

1 Open the Administration window from the Server menu pop-up.

2 Click Response from the panel on the left, and click the Create button to make a new response.

3 From the Response Action pop-up, choose "Restore from Archive." For Name, enter *Restore from Archive*, and for Description enter *Restore primary representation from archive.* Click Save Changes when finished.

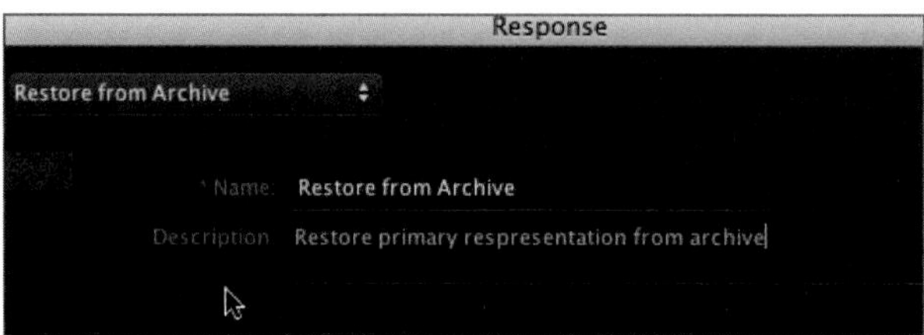

4 Back in the Administration window, choose Subscription from the pane on the left and click the Create button to make a new subscription.

5 From the "Subscribe to" pop-up, choose Asset. For Name, enter *Restore from Archive on Ready for Edit*, and select the Enabled checkbox. For Description, enter *Restore from Archive on Ready for Edit.* Keep Event Priority Modifier at Normal, and choose Modified from the Event Type Filter list. Choose "Restore from Archive" from the Available list, and click Add to move it to the Selected list.

6 Click Asset Filter in the pane on the left.

This is where you'll enter the matching criteria to trigger your subscription.

7 Click the Status pop-up menu and choose Equals, type *Ready for Editing* in the text-entry box next to it, and select the "Trigger if changed" checkbox.

8 Click the Archive Status pop-up menu and choose Equals, and then choose Offline from the next pop-up menu.

Now this subscription will only be triggered when someone changes the Status pop-up menu of an asset to "Ready for Editing," and only if the asset's primary representation is offline.

9 Click Save Changes when finished.

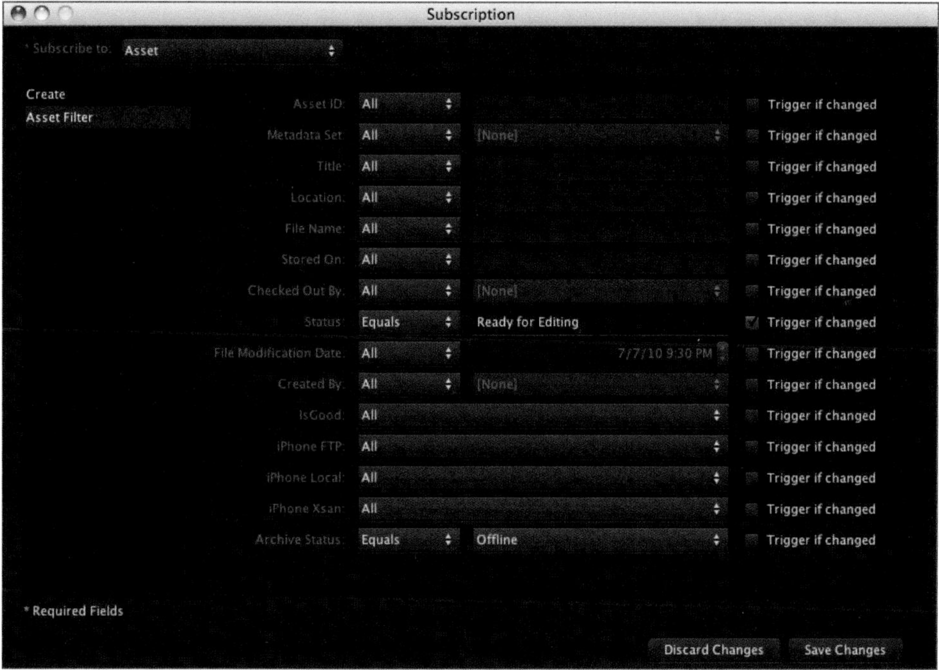

10 Go back to the main client application window. Double-click the **SR1015 002 Copper River Delta-v** asset that you archived using a subscription earlier in this section.

11 Click the Review & Approve tab. From the Status pop-up menu, choose Ready for Editing, and click Save Changes.

12 Open the Jobs window from the main application pop-up menu and observe the "Restore from Archive" job. Depending on the size and location, this process can take some time.

13 When the job has finished, go back to the main client application screen and click Refresh in the upper-right corner. Notice that the **SR1015 002 Copper River Delta-v** asset no longer has the archive icon in Thumbnails view.

You now have a general knowledge of how to follow a streamlined archiving workflow, but more than likely you will need to modify it for your specific hardware/software.

One way this is achieved is through pre- and post-archive script responses. Third-party providers and individuals have posted some great examples online that show integration between Final Cut Server and archival software/hardware. Here are some excellent examples of using scripts to integrate with third-party archive systems:

- Matt Geller's post on Xsanity regarding BakBone NetVault integration:

  www.xsanity.com/article.php/2009040918341824/print

- Andre Aulich's post on his website regarding Archiware PresSTORE integration:

  www.andre-aulich.de/en/perm/connecting-final-cut-server-to-archiware-presstore-concept-overview

- MatrixStore's integration scripts and production archival scripts from Object Matrix:

  www.matrixstore.net/2009/07/24/final-cut-server-15-plugins/

- Atempo's ADA digital archive solution:

  www.atempo.com

Lesson Review

1. What stays online and what is taken offline during the archive process?
2. True or false: Final Cut Server can archive to only one device.

3. Which metadata field can you use to filter whether an asset is offline or online in the Advanced Search?
4. Where can you view the updated status of an archive or restore response?

Answers

1. The proxies and metadata stay online, and the primary representation is taken offline.
2. False. Final Cut Server has the ability to archive to multiple devices.
3. Archive Status.
4. The Jobs window or the Log tab in the Administration window.

Goals

- Apply production scans
- Use the command-line client
- Change Thumbnails, List, and Tile views

Lesson **11**

Advanced Server Scenarios

ADMIN ► This lesson is primarily for the admin, although some users will also find the information useful.

This chapter is dedicated to scenarios and subjects that fall outside the normal deployment of Final Cut Server but can be utilized to extend its flexibility and functionality. Some of these are performance enhancements to accelerate Final Cut project analysis and speed up database access; others deal with automated ways of creating productions from a file-system hierarchy.

Backup is a critical part of your Final Cut Server deployment. The steps to configure backup that will fit your specific workflow are outlined in the *Final Cut Server Setup* and *Administration Guides* provided by Apple.

Production Scans

Production scans are a special type of response in Final Cut Server that allow you to import a directory of files as a new production container with the newly created assets inside. In the Production Scan settings, you specify a directory level at which to create the production and, using some Unix regular expressions, you can map in the directory name as metadata. This comes in handy when you create a lengthy directory naming convention, identifying episode numbers or producers, or the month the content was shot in. All of this information can be mapped to different production metadata fields.

In this task you'll create an example directory structure with some assets underneath it, which you will import directly into Final Cut Server by using a production scan on a schedule.

1 Log in as administrator to your system and launch Finder. Navigate to the /FCSvr directory.

2 Create a new folder at that level called *Projects*. Create three new subfolders inside it: *May_Smith_2008*, *September_Carter_2003*, and *November_Walsh_2006*.

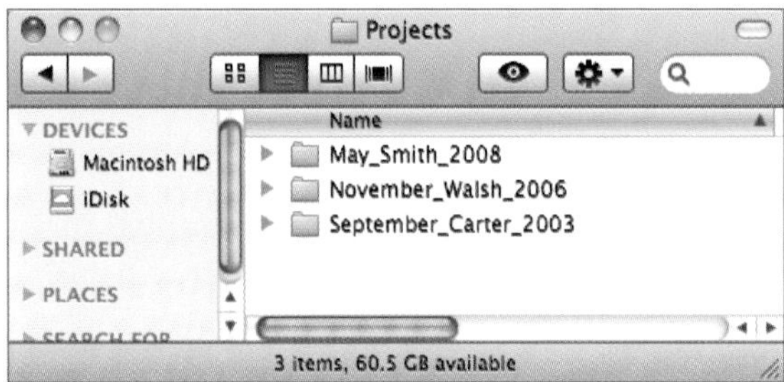

3 To show the content that can be easily inserted into these productions, open another Finder window and navigate to the /Library/Desktop Pictures folder (where OS X stores all of its default desktop art). Option-drag **Aqua Blue.jpg** to the May_Smith_2008 folder, Option-drag **Aqua Graphite.jpg** to the November_Walsh_2006 folder, and Option-drag **Flow 1.jpg** to the September_Carter_2003 folder.

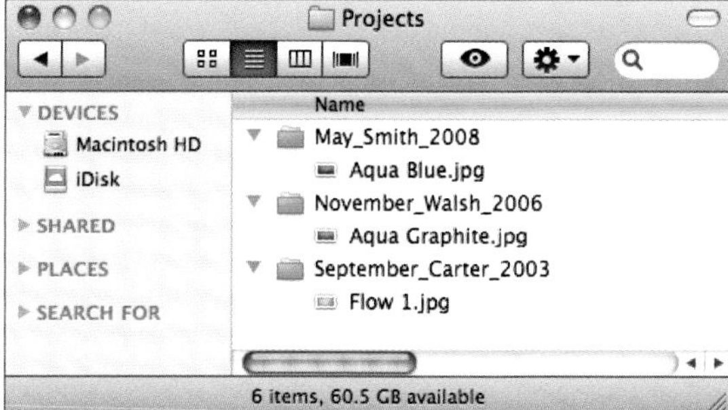

4 Launch System Preferences, click the Final Cut Server icon, and authenticate. Click Devices. Click the Add (+) button to create a new device from the folder structure you just created.

5 Choose Local from the Device Setup Assistant and click Continue.

6 For name, enter *Projects*. Click Browse and navigate to the Projects directory and click Choose. Click Continue.

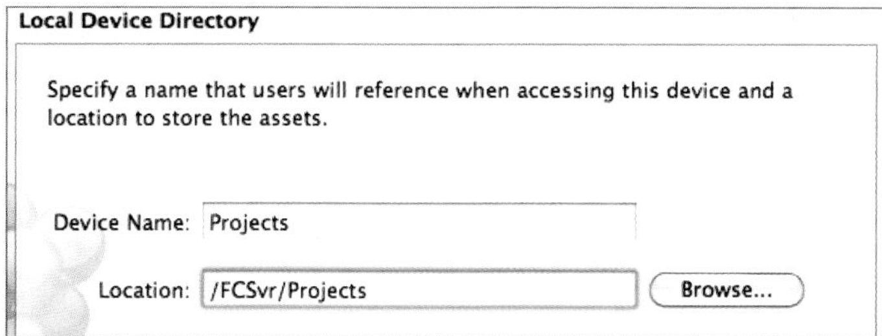

7 You won't be making this an archive device, so leave that option deselected and click Continue. Also, you'll be setting up production scans for this device (which is not an available option in the System Preferences pane), so leave Scan deselected and click Continue.

8 You won't be publishing any content to this device, so leave the default Transcode Settings selected, and click Continue. On the final screen of the Setup Assistant, make sure your settings look like those in the following image, and then click Done to finish creating your new device.

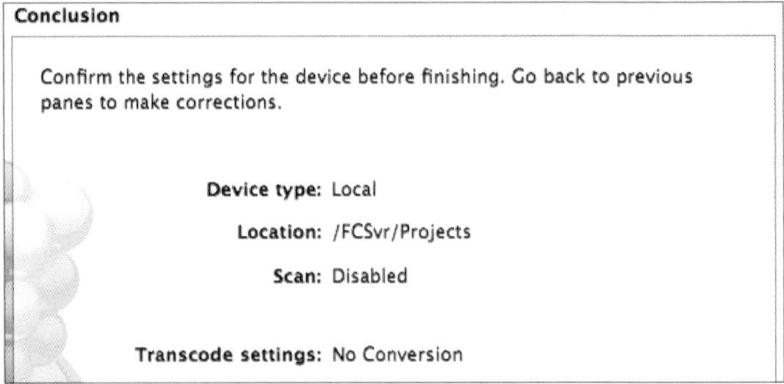

9 Launch the client application and log in as the administrator. Open the Administration window from the Server pop-up menu.

10 Select Response from the pane on the left and click the Create button to make a new response. From the Response Action pop-up menu, choose Scan Productions.

11 Select Create from the pane on the left, and for Name and Description enter *Production Scan of Projects*, and then click Production Scan on the left.

12 From the Scan Source pop-up menu, choose Projects (the device you created earlier). From the Metadata Set pop-up menu, choose Graphic (since you're using graphic images as samples). From the Production Metadata pop-up menu, choose Package.

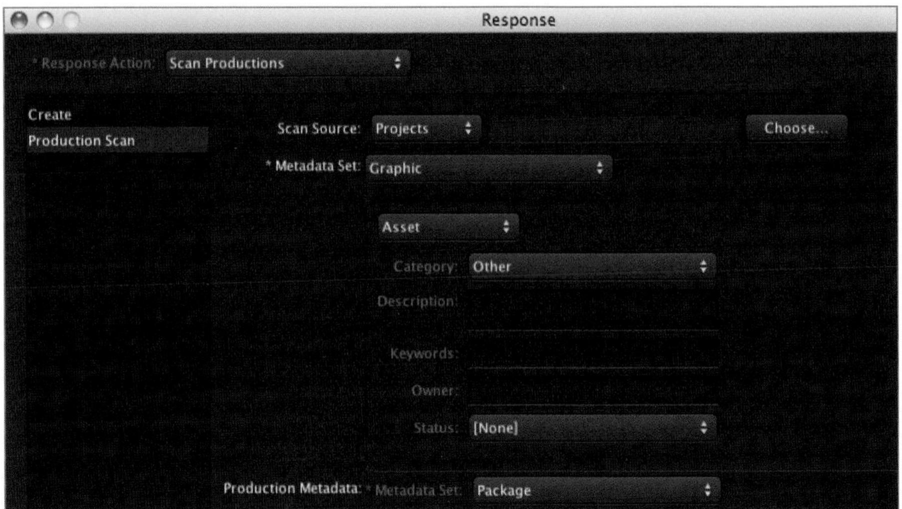

13 In the Title metadata field, enter *[0]*.

This is an example of a regular expression. It tells Final Cut Server to insert the directory title directly into this production metadata field. You can enter in additional metadata if you like, but it's not necessary. (The "Mapping Directory Name Components to Metadata Fields" section looks at regular expressions that are a little more advanced.)

14 For Scan Mode, choose Full; for Entity Type, choose File; and for Production Depth, enter *0*. (This is the number of directories down at which you want to start making productions; you choose 0 because your folders are at the top level of the Projects device.) For Recursion Limit, enter *0*. Leave the other options at default and click Save Changes.

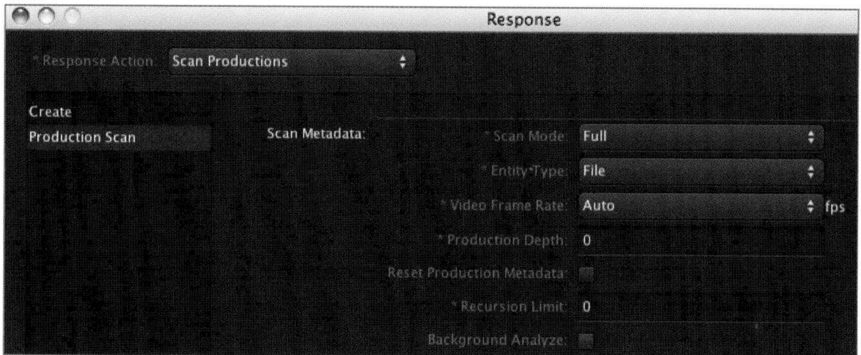

15 Back in the Administration window, in the left pane, click Schedule. Then click the Create button to make a new schedule.

16 From the Schedule pop-up menu, choose Periodically. For Name, enter *Production Scan of Projects*. Select Enabled to turn on the schedule, and enter *Production Scan of Projects* for Description.

17 From the Available list on the right, select Production Scan of Projects and click Add to move it to the Selected list on the left. In the pane on the left, select Schedule Period and enter *2*. This will make your Production Scan response run every two minutes. Click Save Changes.

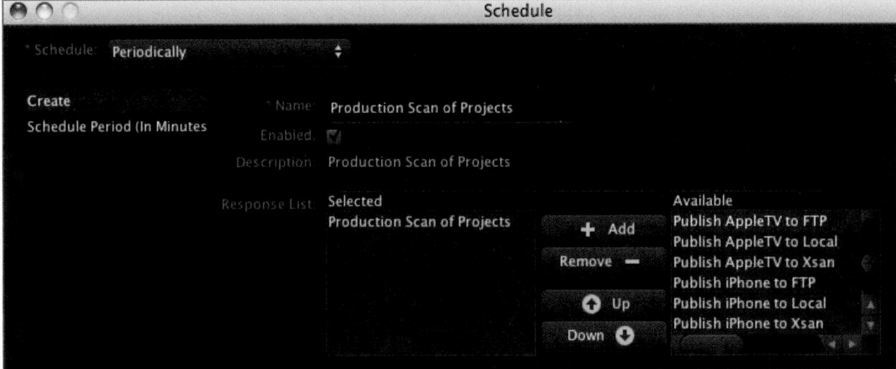

18 Monitor the Search All Jobs window to see your production scan kick off. After two minutes, you'll see the proxy creation process start, and each asset will have all of its representations created.

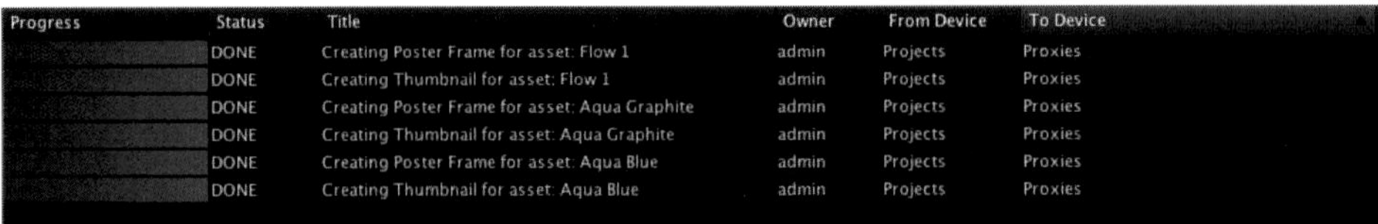

Progress	Status	Title	Owner	From Device	To Device
	DONE	Creating Poster Frame for asset: Flow 1	admin	Projects	Proxies
	DONE	Creating Thumbnail for asset: Flow 1	admin	Projects	Proxies
	DONE	Creating Poster Frame for asset: Aqua Graphite	admin	Projects	Proxies
	DONE	Creating Thumbnail for asset: Aqua Graphite	admin	Projects	Proxies
	DONE	Creating Poster Frame for asset: Aqua Blue	admin	Projects	Proxies
	DONE	Creating Thumbnail for asset: Aqua Blue	admin	Projects	Proxies

19 Back in the Application window, in the upper-left corner, click Productions. Final Cut Server will automatically perform a search.

NOTE ▶ Your new productions were automatically created by the Production Scan response. You'll also notice that the Title fields of the production containers were automatically mapped from the names of the directories you created at the beginning of this exercise. These mappings were the result of placing the *[0]* regular expression in the Title field of your Production Scan of Projects response.

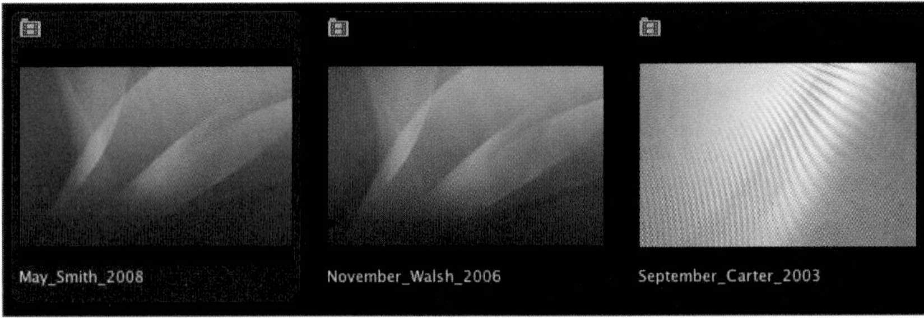

Mapping Directory Name Components to Metadata Fields

Now that you've created a simple example of a production scan with directory mapping, let's try using more complex regular expressions to map specific bits of a directory name into separate metadata fields. You can use these examples as a building block for designing a naming convention, or as a foundation for building out more complex regular expressions.

1 In the left pane of the Administration window, select Response, and then double-click Production Scan of Projects to edit the response you made earlier.

2 Select Production Scan from the left pane, and then scroll down to the Production Metadata section. Enter the following values:

   Title = [0]/^([^_]*)_/

   This will map the beginning value from your directory structure—for example, May.

   Client = [0]/_(.*)_/

   This will map the middle value from your directory structure—for example, Smith.

   Product = [0]/_([^_]*)$/

   This will map the last value from your directory structure—for example, 2008.

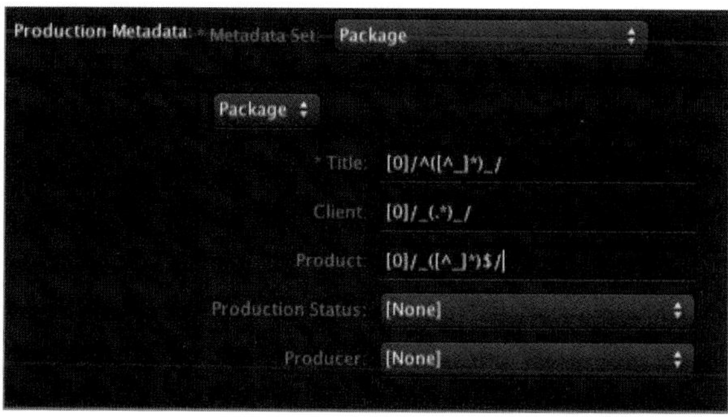

3 In the Scan Metadata section, select the Reset Production Metadata checkbox. This forces existing productions to have their metadata remapped the next time the scan is

run. This is useful when testing out regular expressions as you will likely map metadata incorrectly the first couple of times out. Click Save Changes and click Overwrite in the warning dialog.

In two minutes your Scan Productions response will run again, and because you selected Reset Production Metadata when you added the new regular expressions, your production metadata will be updated.

4 To verify this update, refresh the Productions screen in the client application (after waiting two minutes) and you'll see that the titles of the productions have been updated.

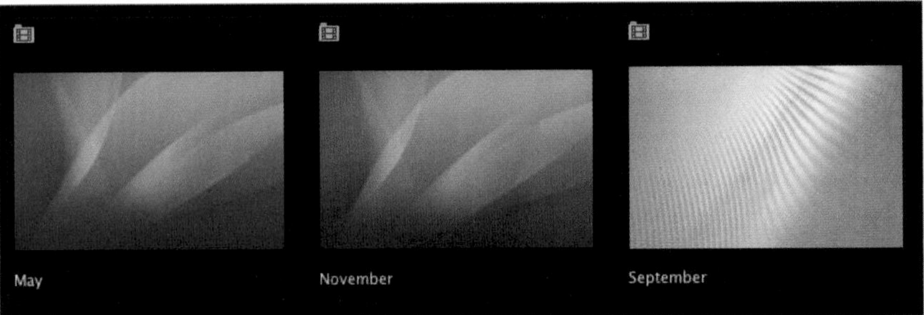

5 Double-click the May production and click the Metadata tab. The Client and Product fields have been automatically updated with the information from the directory name.

You should now have a fundamental understanding of how production scans work, and you can use this information to meld the scans into your workflow. One application for this process would be to use production scans to automatically create productions out of FCP projects in the Capture Scratch folder.

Changing the Default Views

By default, Final Cut Server displays a template of metadata when you're searching for assets and viewing thumbnails, in the List View and in the Tile pane. Like the Asset Filter, these views can be customized by editing special metadata groups that are specific to each function. The point of this exercise is to allow you to display specific metadata in different views. Perhaps you would like to add File Size and Last Modified metadata to the Thumbnails view. The advantage here is that users can see this information directly, so they don't have to double-click the asset to view the metadata.

In this exercise, you'll add the Size and Last Modified metadata fields to the default Thumbnails view and the Last Accessed metadata field to the Tile and List views. You can add any custom metadata fields you create to these groups as well.

1. In the Administration window, click Metadata Group in the pane on the left and do a search for *Thumbnail*. Notice that many results are returned, each with different internal functions. Double-click the Thumbnails group with the Metadata Group ID of ASSET_THUMBNAILS_VIEW.

   Notice that Title is the only field currently selected. This is what is displayed below the thumbnail in the main client application.

2. From the Available list on the right, select Size and Last Modified and click the Add button to move them to the Selected list on the left. In the Selected box, select Size to expose the Field Properties. Select the Scale Numbers option to show the value in KB, MB, and so forth instead of absolute numeric values. Click Save Changes.

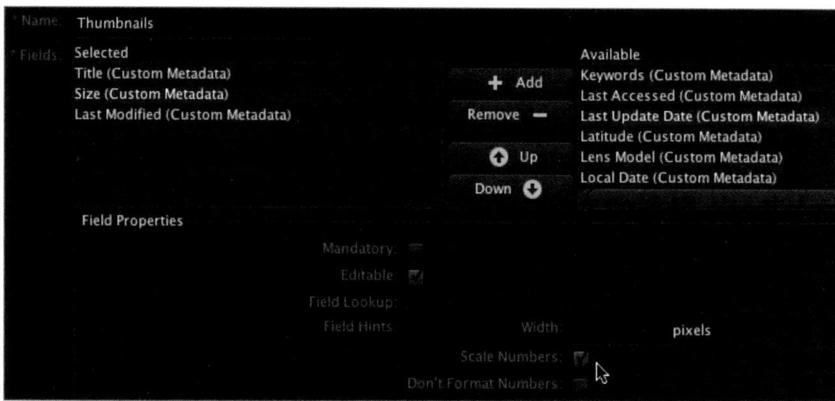

3 Quit the client application, reopen it, and log in as an administrator. Notice that the metadata displayed below the assets in the Thumbnails view has been updated with the Size and Last Modified metadata fields.

Now that you've modified the Thumbnails view, let's add the Status view to the List view.

4 Back in the Administration window, select Metadata Group on the left and do a search for *List*. Double-click the entry with the Metadata Group ID of ASSET_LIST_VIEW to edit the default List view.

5 From the Available list on the right, select Last Accessed and click the Add button to move it to the Selected list on the left. Select the Last Accessed field in the list of Selected fields and click the Up button once to move the field between File Modification Date and Last Modified. Click Save Changes.

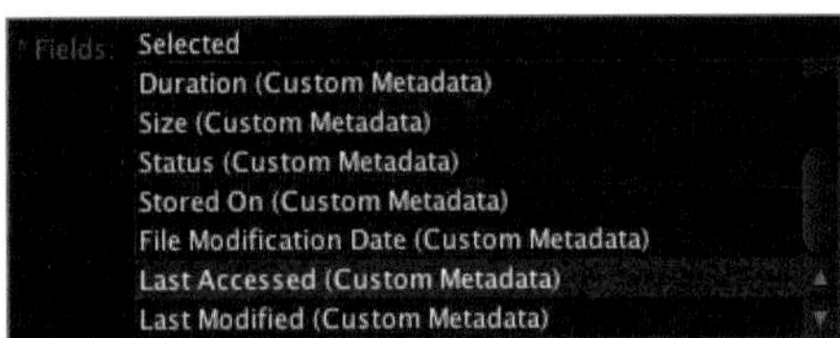

6 Quit the client application, reopen it, and log in as an administrator to see the updated changes to the List view. Notice that the Last Accessed field is now one of the viewable options.

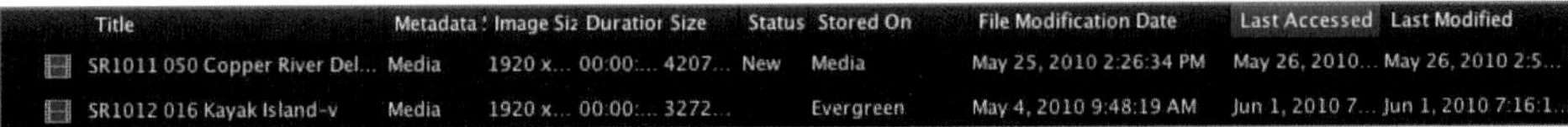

Let's add Last Accessed to the Tiles information view in the lower-left corner of the main Asset screen as well. Like the Thumbnails and List views, the Tiles view is controlled by a special metadata group.

7 In the left pane of the Administration window, select Metadata Group, and do a search for *Tiles*. Double-click the Tiles entry that has the Metadata Group ID of ASSET_INFO_VIEW to edit it.

8 From the list of Available fields on the right, select Last Accessed and click the Add button to move it to the list of Selected fields on the left. Select the Last Accessed field in the list of Selected fields and click the Up button once to move the field above Last Modified. With the Last Accessed field still selected, deselect the Editable checkbox. Click Save Changes.

9 Quit the client application, reopen it, and log in as an administrator to see the updated changes to the Tiles view. Select the **Flow1.jpg** asset and notice that Last Accessed has been added.

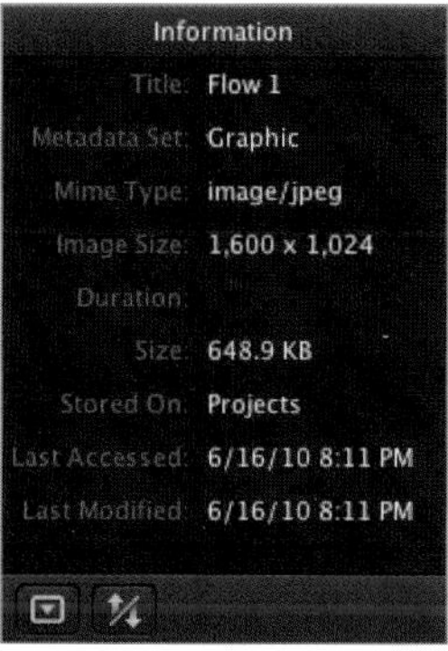

Optimizing the Database

Final Cut Server uses the PostgreSQL open source database as its back-end engine. When it comes to optimizing the database, the number one contributing factor is RAM: The larger your database, the more RAM you should have in your system. As a rough estimate, you can expect 250,000 assets to take up about 6 GB of disk space. As a best practice, you want to have more RAM in your system than the physical size of your database so that the database can be accessed directly from RAM, and not have to swap from the hard drive, which will drastically degrade performance. One of the first symptoms of a lack of sufficient RAM in the system is a search taking longer than a couple of minutes.

At the time of writing, Apple is shipping the Nehalem architecture in its Xserve and Mac Pro computers, which utilize DDR3 RAM. This type of RAM is optimal in three pairs of DIMMs, so make sure to spec out your systems properly to get the maximum amount of memory bandwidth.

Another instance where you would want to change the database to utilize more RAM is if your workflow requires you to have a large number of clips in Final Cut Pro projects—for example, more than 500. When an FCP project is analyzed, linking all the information between assets and elements can be a memory-intensive process. If you're seeing lengthy check-in times with your Final Cut Pro projects due to the number of clips they contain, it's highly recommended that you optimize your database per the recommendations laid out in this section.

It's also possible to modify the configuration of the database to make it access more RAM than it does by default. You have to take many factors into consideration when you decide to go down this path. First, the system can only be running Final Cut Server; no other services can be running, as they will have memory needs as well. Second, the server cannot be a Qmaster node or a Qmaster controller. Again, this is because Final Cut Server is pulling memory away from these other services, which would cause degraded performance on both sides.

You will now examine the default PostgreSQL configuration file and identify the parameters you can modify to change the amount of memory the database can access.

> **NOTE** ▶ The following steps should be undertaken only by an admin who is comfortable in a command-line environment.

1 Log in to the server running the Final Cut Server application as administrator, and open Terminal.

2 Give yourself a root shell by running sudo –s, and authenticating with your admin password.

3 Change your working directory to the location of the database by running the command
cd /var/db/FinalCutServer/data/

TIP You can verify the command by entering the command pwd, which will return your current location.

```
logalog:~ drew$ sudo -s
bash-3.2# cd /var/db/FinalCutServer/data/
bash-3.2# pwd
/var/db/FinalCutServer/data
bash-3.2#
```

4 Before you make any changes to the postgresql.conf file, back it up first. Run the command cp postgresql.conf postgresql.conf.save.

```
bash-3.2# cp postgresql.conf postgresql.conf.save
bash-3.2# ls |grep postgresql.conf
postgresql.conf
postgresql.conf.save
bash-3.2#
```

5 Open postgresql.conf using your favorite command-line text editor. Novices will be most comfortable using nano; more experienced users should be using vi or emacs. To open the file with nano, type nano postgresql.conf.

6 The portions that you should care about pertain to memory usage. You can jump to this portion quickly by doing a search for *shared_buffers*. Note the default memory settings in the following screen shot. The ones you should care about are shared_buffers, work_mem, and maintenance_work_mem.

```
# - Memory -

shared_buffers = 64MB                   # min 128kB or max_connections*16kB
                                        # (change requires restart)
#temp_buffers = 8MB                     # min 800kB
#max_prepared_transactions = 5          # can be 0 or more
                                        # (change requires restart)
# Note:  Increasing max_prepared_transactions costs ~600 bytes of shared memory
# per transaction slot, plus lock space (see max_locks_per_transaction).
work_mem = 128MB                        # min 64kB
maintenance_work_mem = 256MB            # min 1MB
#max_stack_depth = 2MB                  # min 100kB
```

7 A safe starting point is doubling these values, so for shared_buffers, type *128MB*; for work_mem, type *256MB*; and for maintenance_work_mem, type *512MB*.

NOTE ▶ When changing these values, remember that your server still needs RAM to run system and OS processes, so care is needed so that you don't grind your system to a halt.

8 After updating the values, save the postgresql.conf file and quit your text editor. Restart Final Cut Server through the System Preferences pane and test out your new values by comparing them to previous check-in times or search times.

Using Command-Line Binaries

Final Cut Server ships with a command-line client and a number of command-line binaries in the app bundle that provide specific functions. The fcsvr_client binary allows you to communicate with your server and do almost anything that the client application can do. Since it's a binary, you can automate some of your processes through scripting. For integrators, the command-line client can be especially useful, as you can script and package deployments, saving large amounts of time.

In this section, you'll see some examples of using fcsvr_client to add assets to the catalog, to search for assets, to delete assets, and other options. There is some example documentation if you run fcsvr_client without any arguments, which will show you how Final Cut Server handles internal addressing and what some of your options are when using the client.

First, let's look at using fcsvr_client to search for assets that already exist in Final Cut Server. Using fcsvr_client you can find assets based on searchable criteria, same as the client application, but once you get the Asset ID for the asset you want you can output the metadata as XML or parse it into another system. Using this Asset ID, you can also set specific metadata on individual assets.

1 In Terminal, run the following command: `cd "/Library/Application Support/Final Cut Server/Final Cut Server.bundle/Contents/MacOS"`.

   This is where fcsvr_client is stored. If you're going to be using this client often, it will save you a lot of time by adding this location to your $PATH variable, depending on what shell you are using.

2 Try to find **Flow1.jpg**, which you ingested earlier in this chapter. Run the following command: `./fcsvr_client search /asset`.

   This will show you *all* of the assets currently in the database. If you have a lot, this command may take a while to run.

3 To narrow down your search criteria, run the following command: ./fcsvr_client search --crit Flow /asset.

Note that one entry is returned: the Flow 1.jpg image asset that you ingested using your production scan. The most important bit of data that this command reveals is the Asset ID, which in this case is /asset/200.

NOTE ► Your Asset ID may be different depending on the order in which you have worked through the book.

```
logalog:~ drew$ cd /Library/Application\ Support/Final\ Cut\ Server/Final\ Cut\ Server.bundle/Contents/MacOS/
logalog:MacOS drew$ pwd
/Library/Application Support/Final Cut Server/Final Cut Server.bundle/Contents/MacOS
logalog:MacOS drew$ ./fcsvr_client search --crit Flow /asset
Address         ASSET_MEDIA_TYPE            CUST_TITLE                  ASSET_TYPE                  ASSET_IMAGE_SIZ
E               ASSET_DURATION              CUST_SIZE                   CUST_ASSET_STATUS           CUST_DEVICE
                CUST_CREATED                LAST_ACCESSED               MD_LAST_MODIFIED

/asset/200      image                       Flow 1                      pa_asset_media              { x 1600, y 102
4 }             -                           531486                      -                           Projects
                2008-07-29 23:04:11+0       2010-03-10 03:45:37+0       2010-03-10 03:45:37+0
logalog:MacOS drew$
```

4 Now that you know the Asset ID, you can dump the entire record of metadata for that asset by using fcsvr_client. Run the following command: ./fcsvr_client getmd /asset/200.

NOTE ► Replace 200 with the Asset ID from the results in Step 3.

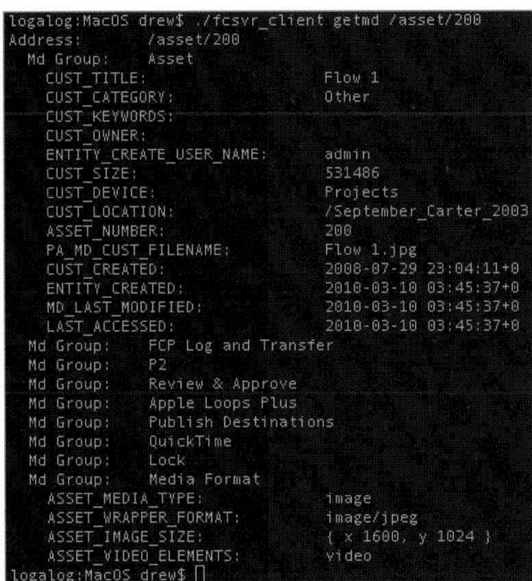

```
logalog:MacOS drew$ ./fcsvr_client getmd /asset/200
Address:        /asset/200
  Md Group:     Asset
    CUST_TITLE:                     Flow 1
    CUST_CATEGORY:                  Other
    CUST_KEYWORDS:
    CUST_OWNER:
    ENTITY_CREATE_USER_NAME:        admin
    CUST_SIZE:                      531486
    CUST_DEVICE:                    Projects
    CUST_LOCATION:                  /September_Carter_2003
    ASSET_NUMBER:                   200
    PA_MD_CUST_FILENAME:            Flow 1.jpg
    CUST_CREATED:                   2008-07-29 23:04:11+0
    ENTITY_CREATED:                 2010-03-10 03:45:37+0
    MD_LAST_MODIFIED:               2010-03-10 03:45:37+0
    LAST_ACCESSED:                  2010-03-10 03:45:37+0
  Md Group:     FCP Log and Transfer
  Md Group:     P2
  Md Group:     Review & Approve
  Md Group:     Apple Loops Plus
  Md Group:     Publish Destinations
  Md Group:     QuickTime
  Md Group:     Lock
  Md Group:     Media Format
    ASSET_MEDIA_TYPE:               image
    ASSET_WRAPPER_FORMAT:           image/jpeg
    ASSET_IMAGE_SIZE:               { x 1600, y 1024 }
    ASSET_VIDEO_ELEMENTS:           video
logalog:MacOS drew$
```

5 Note that the command in the previous step only gave you a subset of metadata information. If you want to see *all* of the metadata for a given asset, you can have this information outputted as XML. To do so, run the following command:

```
./fcsvr_client getmd /asset/200 --xml
```

6 You can also have this information written directly to a file instead of having it outputted to the screen. To do so, run the following command:

```
./fcsvr_client getmd /asset/200 –xml > ~/Desktop/asset200.xml
```

Now that you've seen how to get the Asset ID for an asset through fcsvr_client and how to get metadata information using this Asset ID, let's look at setting metadata from the command line.

7 Using the same asset you've been using as an example, run the following command:

```
sudo ./fcsvr_client setmd /asset/200 CUST_DESCRIPTION="this is a test of the emergency broadcast system"
```

NOTE ▶ This command requires sudo because fcsvr_client needs to access a domain socket to set the metadata.

8 Switch back to the client application. If you closed it previously, reopen it from the desktop icon and log in as an administrator. Do a search for *Flow*. Double-click the **Flow 1.jpg** asset, and note that the metadata you set from the command line has been automatically imported into the asset you specified.

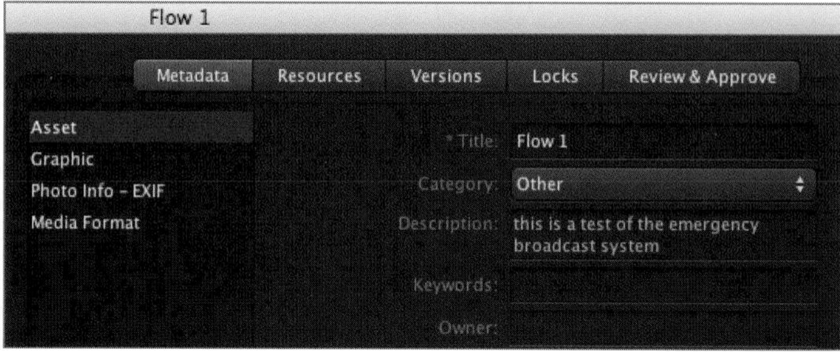

Advanced Command-Line Tasks

Now that you've looked at some of the basic features of fcsvr_client, let's look at some of the more advanced things you can accomplish using fcsvr_client. Sometimes, an automation can become "stuck" and thus prevent any other automations (watchers or subscriptions) from working. This is because Final Cut Server queues these automation events in the database, and they are fired off sequentially. So if one gets stuck, it blocks the execution of others. To remedy this, you can run the following command: `sudo fcsvr_client flush_response_queue`.

Note that if you have other events in the queue, they will be removed as well.

Lesson Review

1. What happens when you select the Reset Production Metadata checkbox?
2. What are two applications for mapping specific bits of a directory name into separate metadata fields?
3. What is the most important contributing factor to optimization of large databases?
4. What is the advantage of using fcsvr_client binaries instead of the client application when searching assets?
5. How do you get access to all of the metadata for a given asset when working with fcsvr_client?

Answers

1. Existing productions are forced to have their metadata remapped and updated accordingly the next time a production scan is run.
2. It can be used as a building block for designing a naming convention, or as a foundation for building out more complex regular expressions.
3. Having adequate system RAM.
4. Once you get the Asset ID for the asset you want you can output the metadata as XML or parse it into another system.
5. Export the information outputted as XML.

Glossary

A

Add to Cache The process of downloading an asset's primary representation and placing it in the cache on your hard drive. This file then stays in the cache, and when you drag and drop the asset using Final Cut Server's interface, a link for that file's location is passed to the location or application to which you drag the file.

analyze A process, available only to administrators of Final Cut Server, that re-creates an asset's other representations, using the primary representation. This is usually done because the primary representation was changed outside of Final Cut Server, or because other representations were not created or are missing.

annotation Timecode-based comments on a video-clip asset that can be viewed by others in your organization. Annotations usually refer to particular sections of the clip, specified by In and Out points.

asset A listing within the Final Cut Server catalog for a file that exists within one of its devices. The file may be a variety of file types including media files, documents, or a folder saved as a bundle. Metadata should be assigned to each asset to ensure fast and targeted catalog searches.

C

catalog The entire database of Final Cut Server, specifically referring to all of the assets and productions that have been created.

check-in The process of returning an asset's primary representation to the Final Cut Server catalog after being checked out and then modified externally. If version control is enabled for the asset, comments on the new version are recorded by the person checking the asset back in, and the old version of the asset is saved in the Version device. Once checked in, the asset's lock is removed and others may modify its metadata or check it out.

checkout The process of both locking and exporting an asset's primary representation for the purposes of external modification. While an asset is checked out, other users are prohibited from modifying the asset's metadata or other representations until the asset is checked in.

clip proxy A video clip that Final Cut Server makes from the primary representation of a video asset. It is much smaller and of lower quality than the primary representation, and as a result, it is easier to transmit over Ethernet networks.

codec Literally, COmpressor/DECompressor. A mathematical algorithm used to make a video or audio clip smaller, while maintaining a predetermined level of quality.

Common Internet File System (CIFS) Network protocol for a Windows-based file server.

composite clip Two files found within the same folder of a device: an RGB clip and its corresponding, separate alpha channel clip. Both files usually have the same name before their file extension.

contentbase A device used exclusively by Final Cut Server to store files for specific tasks like proxies and version control. Contentbases are formed as bundles to protect their contents from casual interaction in the Finder.

D

DNS (Domain Name System) A service provided on an Ethernet network that resolves IP address numbers (such as 192.168.1.20) to domain names (such as fcserver.pretendco.com). A well-implemented DNS system provides these resolutions in both forward (name to number) and reverse (number to name) directions, and also passes questions it can't resolve to a larger DNS server elsewhere in the organization, or perhaps the Internet itself.

duplicate The process of copying the primary representation of an asset from one device to another (or back onto the same device, usually changing the name to differentiate it from the file that exists). The option to transcode the resulting file is available.

E

edit-in-place device A device that may be directly accessible to users using the client application. If it is, Final Cut Server can provide a path to a file, given in the form of a URI, rather than the file having to be downloaded to the user's cache.

edit proxy A representation of a video asset, used for the purposes of editing clips.

export The process of downloading the primary representation of an asset to a specific location on your computer using the client application. You can opt to transcode the primary representation during this process, so that the file that is received by your computer is in a different format.

F

File Transfer Protocol (FTP) A common network protocol for a file server, used primarily for sharing files on the Internet. FTP servers are therefore commonly accessed from remote locations.

Final Cut Server The Mac that is running installed and configured Final Cut Server software. It may also refer to the interaction between the Final Cut Server and the client application in general.

Final Cut Server software The software that gets installed and configured on the Mac that will become the Final Cut Server.

G

group A collection of users. In Final Cut Server, groups have certain access and functionality privileges that are assigned to them through the association with a permission set.

H

Hierarchical Storage Management system (HSM) A storage system designed for backup and archive. HSMs can continuously accept files into their file system and then, according to predefined rules, transfer the data essence of the file to more stable storage media, like tape, even though the file will still appear to be on the original file system. When a user or process wants to restore the file, it will appear that the file is there, even though its data essence is being retrieved from tape.

L

Library A filesystem device, created by Final Cut Server during initial installation, intended to store commonly accessed files, such as stock media and logos.

lock The process that prevents an asset's representations or metadata from being modified or overwritten by any other user. Administrators can cancel locks.

lookups Another way of saying pop-up menus or pick lists. Lookups are assigned to metadata fields in order to control the choices that a user has in filling out a field.

M

metadata Data about data. A standardized model for including searchable information for a file.

metadata field The basic storage unit for a piece of metadata in Final Cut Server.

metadata group A collection of metadata fields, assembled to describe a feature of an asset or production. Specialized metadata groups are also used in Final Cut Server to describe the fields shown in search results, as well as for filtering criteria used during advanced searches.

metadata set A collection of metadata groups that defines all of the metadata for any asset, production, or job within Final Cut Server.

N

Network File System (NFS) A network file server protocol used in many Unix/Linux environments.

Open Directory Apple's marketing term for a collection of directory system binding and communication protocols, which includes its implementation of the open-source Lightweight Directory Access Protocol (LDAP), used as a centralized directory and authentication system. It also includes plug-ins to interact with other directory systems, including Active Directory.

O

permission set A combination of functionality and access settings for a group.

P

Prepare for Disconnected Use The same process as Add to Cache, with the addition of making an alias of the file located within /Users/username/Documents/Final Cut Server/Media Aliases/. This allows the user to use the primary representation of the asset even when the user is not connected to Final Cut Server.

primary representation The original file that an asset refers to. In most cases, this is the full-resolution clip or high-resolution image that resides on a device.

production A "virtual folder" within Final Cut Server, meant as a means to gather together media assets, project assets, and related document assets within a single container.

proxies device A contentbase device dedicated to storing clip proxies.

Redundant Array of Independent Disks (RAID) A system that combines a series of hard disk drives together in order to provide greater speed than a single drive. It may also employ a redundancy provision that allows one or more of the drives to fail without losing data.

R

representation A file to which an asset is associated. An asset is associated with at least a primary representation. Other representations can be thumbnails, clip proxies, edit proxies, and poster frames.

scheduled event or **schedule** A trigger that occurs in an absolute or periodic time frame. These triggers usually start scans that add and purge media. They also can execute maintenance tasks such as clearing logs.

S

Server Message Block (SMB) A network protocol for a Windows-based file server.

still sequence A collection of images that has some sort of iterative numbering system in the filename. The collection usually describes a moving image, with each file being the next sequential frame.

subscription A trigger that looks for changes in the metadata of an asset, production, or job. When the changes occur, a subscription then activates any number of responses.

transcode The process of converting a file from one format into another, using a specific codec.

T

unlock The process of removing a lock from an asset, making it, its representations, and its metadata available for modification by others.

U

upload The process of copying files from your computer to Final Cut Server using the client application. Files can be uploaded either by dragging them into the client application's window or by selecting Upload from the Server pop-up menu.

Version device A contentbase device dedicated to storing older versions of an asset, used by the version control system in Final Cut Server.

V

watcher A folder that is constantly being "watched" by Final Cut Server for new or modified content, using a poll. Watchers are assigned responses by the administrator, including ones to copy and/or transcode the files placed inside a watcher to other devices.

W

Xsan The marketing term for Apple's Storage Area Network (SAN) system. Xsan allows up to 64 computers to share large volumes of data with real-time access to every file.

X

Xserve Apple's enterprise-class, rack-mountable computer, designed for server applications, such as web server, file server, DNS, and Open Directory. Xserves have field-replaceable redundant drive modules, power supplies, and motherboards. They take up only one rack space (1U) in a standard 19-inch rack.

Index

B – C

D

E

F

Q – R

S

T

U – V

W

X

Apple Certification
Fuel your mind. Reach your potential.

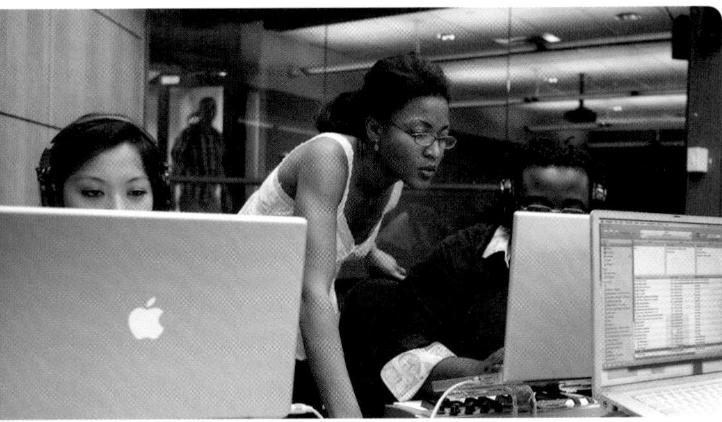

Stand out from the crowd. Differentiate yourself and gain recognition for your expertise by earning Apple Certified Pro status to validate your Final Cut Server skills.

This book prepares you to earn Apple Certified Pro—Final Cut Server Level One. Level One certification attests to essential operational knowledge of the application. Level Two certification demonstrates mastery of advanced features and a deeper understanding of the application. Take it one step further and earn Master Pro certification in Final Cut Studio.

Three Steps to Certification

1. Choose your certification path. More info: training.apple.com/certification.
2. Select a location:

   Apple Authorized Training Centers (AATCs) offer all exams (Mac OS X, Pro Apps, iLife, iWork, and Xsan). AATC locations: training.apple.com/locations

   Prometric Testing Centers (1-888-275-3926) offer all Mac OS X exams, and the Final Cut Server Level One exam. Prometric centers: www.prometric.com/apple
3. Register for and take your exam(s).

"Now when I go out to do corporate videos and I let them know that I'm certified, I get job after job after job."

—Chip McAllister, Final Cut Pro Editor and Winner of The Amazing Race 2004

Reasons to Become an Apple Certified Pro

- **Raise your earning potential.** Studies show that certified professionals can earn more than their non-certified peers.
- **Distinguish yourself from others in your industry.** Proven mastery of an application helps you stand out from the crowd.
- **Display your Apple Certification logo.** Each certification provides a logo to display on business cards, resumes and websites.
- **Publicize your Certifications.** Publish your certifications on the Apple Certified Professionals Registry to connect with schools, clients and employers.

Training Options

Apple's comprehensive curriculum addresses your needs, whether you're an IT or creative professional, educator, or service technician. Hands-on training is available through a worldwide network of Apple Authorized Training Centers (AATCs) or in a self-paced format through the Apple Training Series and Apple Pro Training Series. Learn more about Apple's curriculum and find an AATC near you at training.apple.com.

training.apple.com/certification